Akihisa Inoue
Fanli Kong
Fahad Al-Marzouki

Fe-based Glassy and Nanocrystalline Magnetic Alloys

AF155824

Akihisa Inoue
Fanli Kong
Fahad Al-Marzouki

Fe-based Glassy and Nanocrystalline Magnetic Alloys

LAP LAMBERT Academic Publishing

Publisher:
LAP LAMBERT Academic Publishing
is a trademark of
Dodo Books Indian Ocean Ltd. and OmniScriptum S.R.L publishing group

120 High Road, East Finchley, London, N2 9ED, United Kingdom
Str. Armeneasca 28/1, office 1, Chisinau MD-2012, Republic of Moldova, Europe
Managing Directors: Ieva Konstantinova, Victoria Ursu
info@omniscriptum.com

Printed at: see last page
ISBN: 978-3-659-61735-5

CONTENTS

1. Introduction

Since the first synthesis of an amorphous phase in Au-Si system by rapid quenching of melt in 1960 [1], a great large number of amorphous and glassy alloys were prepared up to date. Their alloy systems cover almost all metallic elements in the periodic table and can be classified to ferrous and non-ferrous alloy groups. Considering the present application situation and its future prospect of amorphous and glassy alloys, ferrous metal-based amorphous and glassy alloys are particularly important and attractive and their engineering importance has steadily increased toward the creation of a low-carbon society. When we look at the development history of Fe-based amorphous and glassy alloys, the first synthesis of Fe-based amorphous alloy was made for Fe-P-C system in 1967 [2], followed by the finding of magnetic softness for Fe-B-C amorphous alloy [3]. An important alloy composition patent which covers (Fe,Co,Ni)-(P,Si)-(B,C) amorphous alloys was applied by Chen and Polk in 1972 and Allied Chemical Corporation succeeded in obtaining the patent license of the Fe-based amorphous alloys with extensively wide alloy compositions in 1974 [4]. In addition, another patent of planar flow casting technique leading to the formation of wide glassy alloy sheets with width above 12 mm was also obtained by Allied Chemical Co. in 1984 [5]. As a result, any other research groups cannot utilize freely their Fe-based amorphous alloys as future commercial materials. In 1985, Allied Chemical Co. succeeded for the first time in commercializing Fe- and Co-Fe-based amorphous alloy sheets as soft magnetic materials for pole transformer cores etc. Much later, Fe-based bulk glassy alloys were synthesized for the first time by copper mold casting method in 1995 [6].

Based on the development and patent situations of Fe-based amorphous alloys in the middle and late 1970s and the early 1980s which were described above, we tried to develop new engineering amorphous alloys except Fe-based amorphous ribbons/sheets developed previously by the planar flow casting method. Here we introduce the development examples of Fe- and Co-based amorphous alloys with unique morphologies such as wire [7, 8], flaky powder [9, 10] and spherical particle [11] and Fe-C base amorphous steels with new alloy compositions [12, 13], in conjunction with their applications.

2. Amorphous Alloy Wires

The Fe- and Co- based amorphous alloy wires with a circular cross section were produced by the in-rotating-water melt spinning method where alloy liquid is injected through a quartz nozzle into a rotating water layer [7, 8, 14]. The resulting Fe-based amorphous wires in Fe-Si-B system have diameters of 100 to 200 μm and exhibit a good ductility which is shown by bending deformation through 180 degrees without fracture. Besides, their wires can be cold drawn to above 80 % reduction in cross section without any intermediate annealing treatment. The cold drawn wires exhibit high a tensile fracture strength of about 3700 MPa and distinct plastic elongation of about 2 % which are much higher than those for conventional piano wire [8]. Tensile fracture of the amorphous alloy wire takes place along the maximum shear stress plane and the fracture surface consists of smooth and vein patterns.

The Fe- and Co-based amorphous alloy wires also exhibit unique and useful soft magnetic properties [15]. The as-spun Fe-Si-B amorphous wires with a diameter around 100 μm exhibit a large ΔE effect exceeding 6%. The cold drawing causes significant increases in AC coercivity and squareness ratio for hysteresis B-H loop through the change in magnetic domain structure. The highest tensile fracture strength and the largest AC coercivity were obtained for $Fe_{75}B_{15}Si_{10}$ amorphous alloy wire of 125 μm in original diameter subjected to cold drawing to about 40 % reduction in cross section, while the Young's modulus decreases gradually with increasing reduction in cross sectional area. Similar behavior was also recognized for $Co_{72.5}B_{15}Si_{12.5}$ amorphous wire with the same original diameter. In addition, the applied field dependence of AC coercivity shows significant difference between the as-spun wires and the cold-drawn wires [15]. As shown in Fig. 1, the as-spun Fe- and Co-based amorphous wires exhibit a linear increase of AC coercivity with increasing applied field. On the other hand, the cold-drawn Fe- and Co-based amorphous wires exhibit nearly constant AC coercivities of about 18 Oe for the Fe-based wire and about 3.5 Oe for the Co-based wire which are less dependent on applied field.

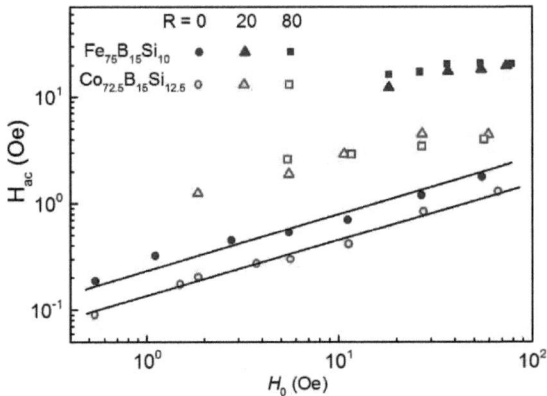

Fig. 1 Applied field dependence of ac coercive force H_{ac} of amorphous $Co_{75}B_{15}Si_{10}$ and $Co_{72.5}B_{15}Si_{12.5}$ wires [15].

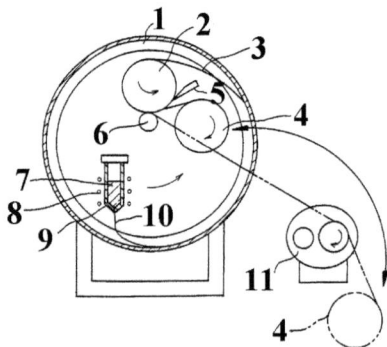

Fig. 2 Schematic illustration of the in-rotating-water spinning apparatus combined with a winder. 1. Rotating Drum; 2. First Magnet Roller; 3. Wire specimen; 4. Second Magnet Roller; 5. Scraper; 6. Nip Roller; 7. Molten Alloy; 8. High Frequency Coil; 9. Melting Crucible; 10. Ejected Alloy; 11. Winder [16].

Based on the magnetic and mechanical properties, dynamic magnetic behavior of the cold-drawn Fe- and Co-based amorphous wires was examined in a pickup coil having 500 turns. Exposure to a switching field of about 200 Oe produced a sharp voltage which was as high as $6V/cm^2$ per turn. It is noticed that flux reversal occurs at very high speed, and hence the wires have been used to generate sharp voltage pulses. By utilizing these novel features, the cold-drawn Fe- and Co-based amorphous wires have been commercialized as the sensing elements in various kinds of devices such as switches, flowmeters, tachometers, proximity sensors, credit cards and robotic sensors etc. Here it is also important to point out that the mass production technique of amorphous alloy wires using the in-rotating-water spinning technique was established by Unitika Corporation in around 1985 [16]. Figure 2 shows a schematic illustration of the established mass production equipment of amorphous alloy wires [16]. The length of the amorphous wires which can be produced in one operation exceeds several hundreds of meter. It is noticed that the equipment is able to wind simultaneously the as-spun long amorphous wires.

By using the same wire production technique, Fe-Si crystalline alloy wires were also developed with the aim of applying to magnetic sensors with much higher saturation magnetization and inexpensiveness. Figure 3 shows optical micrographs revealing the transverse cross sectional structure of melt-spun Fe-Si alloy wires containing 4mass%Si, 6%Si and 7%Si [17]. The as-spun wires have a fine cellular structure for Fe-4mass%Si and Fe-6mass%Si alloys and a dendrite structure for Fe-7mass%Si alloy. The cellular grain size is about 6 μm for the 4%Si wire and about 4 μm for the 6 %Si wire, while the dendrite arm spacing for the as-spun 7%Si wire is in the range from 2 to 6 μm which increases with the distance from surface. Such a change in the dendrite arm spacing can be explained from the theoretical values calculated using a fixed heat transfer coefficient. The Fe-Si wires after annealing for 3.6 ks at 1273 K exhibit good soft magnetic properties of high saturation magnetization exceeding 1.9 T and low coercivity values of 0.8 Oe for the 4%Si wire and 0.6 Oe for the 6%Si wire.

The tensile strength and elongation of the annealed 6%Si wire are about 600 MPa and about 8 %, respectively. The Fe-Si alloy wires have also been tested as sensing elements in AC applied field, but the real commercial stage was not achieved because of the difficulty of developing an inexpensive mass production technique which is suitable for crystalline Fe-Si alloy wires with high melting temperature and high liquid viscosity.

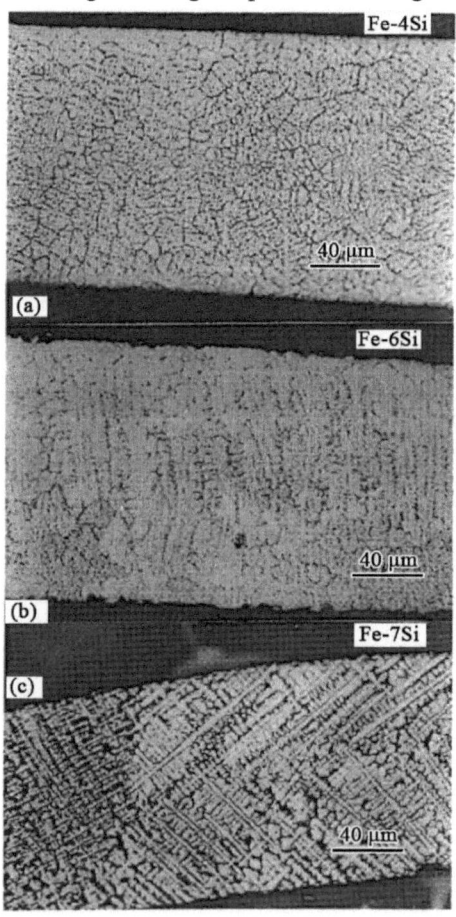

Fig. 3 Optical micrographs revealing the microstructures of the as-spun Fe-Si alloy wires with different Si contents of 4 to 7 mass% [17].

3. Metastable Ordered Austenite Wires

Metastable ordered austenite wires were also developed in Fe-Ni-Al-C [18], Fe-Mn-Al-C [19], Fe-Ni-Cr-Al-C [20] and Fe-Ni-Cr-Al-C-Mo [20] systems by using the same melt-spinning technique. The as-spun wires consist of an ordered austenite structure with grain size of 0.2 to 4 μm and exhibit high yield strength up to about 1700 MPa and distinct plastic elongations. These ordered austenite wires also exhibit a good cold deformability. Table I summarizes static and dynamic mechanical properties and corrosion resistance for the Fe-based ordered austenite wires subjected to cold drawing to about 90 % reduction in cross section, together with the data for commercial piano wire and SUS304 wire. The wires can be cold drawn to 90 % reduction in cross section without any intermediate annealing treatment and the drawn wires exhibit very high tensile fracture strength above 3700 MPa, larger cyclic numbers up to final bending fatigue fracture and better corrosion resistance as compared with piano wire and SUS 304 wire [21]. Consequently, the cold-drawn austenite wires were used as cutting wires for foods such as butter, cheese, fishes and meats etc.

Table I Comparison of properties of melt-spun and 90 % cold drawn Fe-based alloy wires with reference to commercial piano wire and SUS 304 wire [21].

Alloy (at%)	σ_f (MPa)	ε_p (%)	N_f	N_t	H_v (DPN)	R (mg/dm²/d)	V (V/SCE)	D (A/m²)
$Fe_{68}Ni_{20}Si_{10}C_2$	2850	0.5	1.3×10^5	17	580	640	-0.13	0.17
$Fe_{67}Ni_{20}Si_{10}C_3$	2700	0.4	---	13	610	---	---	---
$Fe_{66}Ni_{20}Si_{12}C_2$	3000	0.6	---	15	610	---	---	---
$Fe_{63}Ni_{20}Cr_5Si_{10}C_2$	2800	0.8	1.6×10^5	24	580	580	-0.11	0.14
$Fe_{74}Ni_8Cr_{12.5}Al_{2.5}C_3$	4030	0.3	1.8×10^5	8	725	720	-0.16	0.20
$Fe_{71.5}Ni_8Cr_{10}Al_{7.5}C_3$	3700	0.7	2.1×10^5	15	700	660	-0.14	0.18
$Fe_{74}Ni_8Cr_{12}Al_2C_3Mo_1$	3730	0.7	2.3×10^5	13	700	650	-0.13	0.18
0.8 wt% C Piano wire (d=80 μm)	3380	0.3	4.9×10^4	25	665	5950	-0.46	$>10^3$
SUS 304 (d=100 μm)	980	30	1.0×10^5	67	200	50	-0.03	0.04

σ_f :	Tensile fracture strength	H_v :	Vickers hardness
ε_p :	fracture elongation	R :	weight loss by corrosion
N_f :	number of cycles to failure under the maximum bending strain , λ=0.005	V :	corrosion potential
N_t :	number or twists to failure	D :	anodic current density

4. Flaky Amorphous Powders

By utilizing the significant temperature dependence of viscosity in supercooled liquid below melting temperature for amorphous alloys, amorphous alloy powders with unique outer shape and morphology have been produced for Fe- [9], Co- [10] and Al- [22] based alloy systems. As typical examples, Figure 4 shows SEM images revealing the change in the powder morphology with powder size for $Fe_{78}Si_9B_{13}$ alloy powders produced by a two-stage quenching technique, in which fine supercooled liquid droplets with very high moving velocity generated by high-pressure gas atomization are subjected to impact flattening on a very rapidly rotating rotator [9].

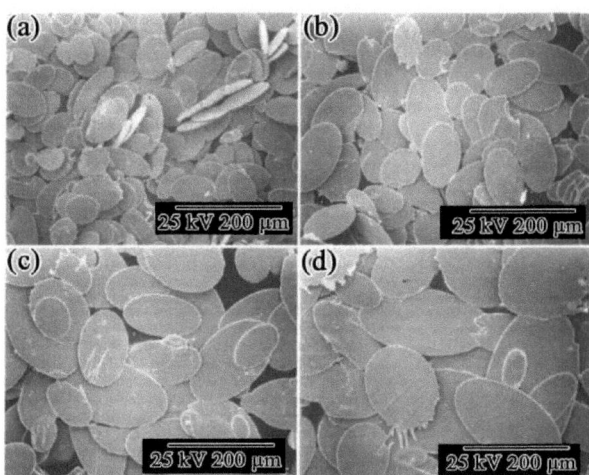

Fig. 4 SEMs showing the effect of size fraction on the morphology of $Fe_{78}Si_9B_{13}$ powders produced by the two-stage quenching method: (a) 25-45 μm; (b) 45-63 μm; (c) 63-90 μm; (d) 90-150 μm [9].

As seen in Fig. 4, the Fe-based flaky amorphous powders have an ellipsoidal shape with very thin thicknesses of 1 to 3 μm and large aspect ratios of 20 to 300 [9]. It is also noticed that the surface of the flaky powder is very smooth and the thickness is uniform. Similar flaky amorphous powders have also been produced for Co-Si-B and Co-Fe-Si-B systems. The composite ring cores consisting of the $Co_{70.5}Fe_{4.5}Si_{10}B_{15}$ flaky amorphous powders embedded in resin exhibit unique soft magnetic properties with very wide maximum and effective permeability (μ_e) ranges which are suitable for application to soft magnetic materials in high frequency fields as well as magnetic sensors. In addition to the usefulness as soft magnetic materials, Fe-based flaky amorphous powders in Fe-Cr-P-C system are also useful as coating materials with features of high corrosion resistance, high hardness and good metallic luster.

5. Spherical Amorphous Particles

Isolated spherical amorphous alloy particles with diameters of 0.5 to 2.0 mm were produced for $Fe_{40}Ni_{40}C_{10}B_{10}$ [11] and $Ni_{64}Pd_{16}P_{20}$ [23] alloys by the injection quenching method to stirred water. The Fe-Ni-based amorphous particles were tested for application to balls for ball point pens. A number of ball point pens using the amorphous alloy balls were made to test the performance ability and endurance limit. However, it was difficult for the ball point pens to keep good soft touch feeling continuously because of high hardness and high stiffness of the glassy balls. As a result, the application to ball point pens was unsuccessful.

Recently, Kawasaki et al. have developed a new type of equipment of producing mono-sized glassy alloy particles by use of actuator [24]. The equipment has unique features such as high yield ratio of approximately 100 %, wide particle size range of 0.1 to 0.5 mm, constant particle size, clean surface and high sphere ratio with size scattering less than 2 %. In addition, the particles have been tested to make 3D printing patterns.

6. Amorphous Alloy Steels

As a completely different type of amorphous alloy, Fe-based amorphous alloys named as amorphous steels were developed by melt spinning of Fe-C-X (X=Cr, Mo, W) ternary, quaternary and pentad alloys [12, 13]. For instance, Fig. 5 shows the composition range of amorphous phase in melt-spun Fe-Cr-Mo-18%C alloys. It is notable that amorphous alloys are formed in a very wide composition range of 0 to 50 at% Cr and 4 to 25 at% Mo. Besides, the C concentration range for formation of an amorphous phase in the Fe-Cr-Mo-C system extends in a range of 15 to 22 at%. The melt-spun amorphous steel ribbons in the Fe-C-X system have high tensile fracture strength of 3450 to 3790 MPa, high corrosion resistance and rather high ductile-brittle transition temperature of about 573 to 673 K after annealing for 100 min.

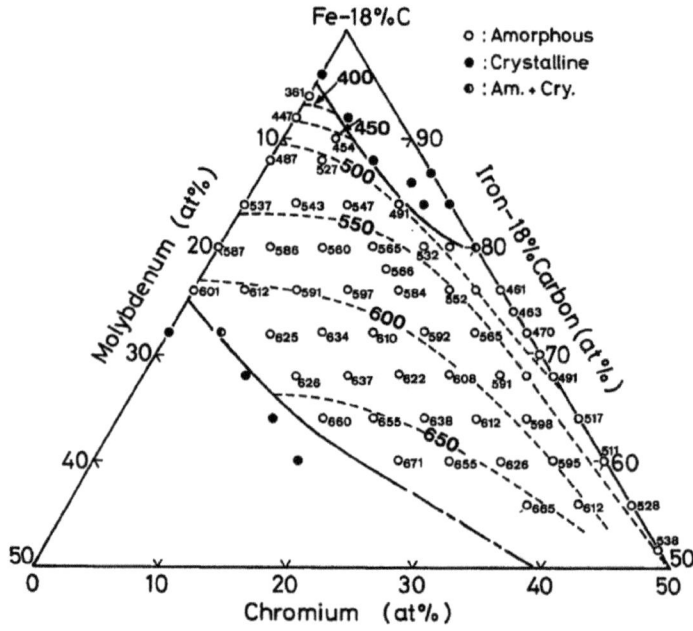

Fig. 5 Composition range for formation of amorphous phase in Fe-Cr-Mo-18%C system and a change in crystallization temperature (T_x) of these alloys [12].

We have also clarified the composition ranges in which various metastable phases are formed in Fe-C-Cr, Fe-C-Mo, Fe-C-W, Fe-C-Al-Ni, Fe-C-Al-Mn and (Fe, Co, Ni)-Mo-C alloys by melt spinning. Their metastable single phases have been reported as amorphous, chai (χ), epsilon (ε), kappa (κ) ordered fcc and fcc phases [26]. By utilizing the nanocrystallization of their metastable single phases, high-elevated temperature strength materials in a bulk compacted form were prepared through the hot-pressing process [25, 27]. As an example, Fig. 6 shows the temperature dependence of Vickers hardness of some Fe-based amorphous steel compacts [25]. It is noticed that the Fe-Cr-Mo-C compact exhibits very high elevated temperature hardness of about 800 at the high temperature of 1073 K, in addition to high oxidation resistance. The amorphous steel compacts were also tested for application to heat resistance tool materials by some companies in Japan and foreign countries.

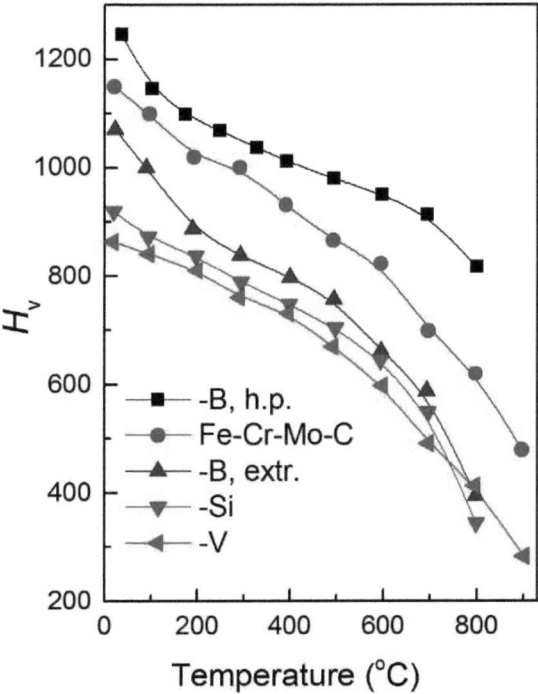

Fig. 6 Elevated temperature hardness of hot-pressed and extruded Fe-16Cr-8Mo-18C-4B, and hot-pressed Fe-16Cr-8Mo-18C-4V alloys. The data for hot-pressed Fe-16Cr-8Mo-18C and Fe-16Cr-8Mo-18C-4Si alloys are included for comparison [25].

7. Engineering Bulk Glassy Alloys

Since around 1990, we can utilize a new type of bulk metallic alloys consisting of glassy structure, in addition to conventional bulk metallic crystalline alloys. The bulk glassy alloys have been used or tested for applications in various fields, e.g., structural, sensor, wear resistance coating, corrosion resistant, magnetic, precise molding, spring, sporting goods, micro- and nano-technology, information data storage, biomedical and fuel cell separator materials in conjunction with net- and near net-shape processings and simple production and working equipment [28, 29]. Among these fields, Fe-based bulk glassy alloys have been used as practical materials in the six fields from structural to precise molding materials in the above-described application lists. Here some typical examples of their applications are presented.

When we look at the development history of Fe-based bulk glassy alloys, their alloys were reported as the following orders; Fe-(Al,Ga)-(P,C,B,Si) [6], Fe-(Al,Ga)-(Cr,Mo,Nb)-(P,C,B,Si) [30], Fe-(Zr,Hf,Nb,Ta)-B [31], Fe-(Cr,Mo,Nb)-(P,C,B,Si) [32], Fe-Ln-B (Ln: lanthanoid) [33], Fe-(C,Si,P,B) [34], Fe-(Cr,Mo)-(C,B) [35], Fe-B-Si-Nb [36] and Fe-(Cr,Mo)-(C,B)-Ln [37, 38]. Here it is important to describe that these Fe-based bulk glassy alloys have been developed by utilizing the three component rules [39, 40], namely, (1) multicomponent alloy systems consisting of more than three group elements; Fe-(Si,P)-(B,C), (2) significant atomic size mismatches above 12 %; Fe > Si,P > B,C, and (3) negative heats of mixing; Fe-Si, Fe-P, Fe-B, Si-B, Si-C, etc.. Besides, we have utilized additional effectiveness of Al, Ga, ETM, LTM and Ln additions for further enhancement of the three component rules. The further multiplication of alloy components also causes the increase in confusion entropy which has been thought to enhance the glass-forming ability.

For instance, the maximum glassy rod diameter in Fe-(P,C,B,Si) base system produced by copper mold casting is about 2.5 mm for Fe-Ga-(P,C,B,Si) [41], 5 mm for Fe-Mo-(P,C,B,Si) [42] and 6 mm for Fe-Co-Mo-(P,C,B,Si) [43]. The rather high glass-forming ability has enabled to produce ring-shaped cores with an outer diameter of 10 mm and thicknesses of 0.5 to 1 mm by copper mold casting. Their bulk glassy alloys also exhibit good combination of rather high B_s above 1.2 T, low coercivity of 1 to 3 A/m, very high maximum permeability reaching about 400,000 and effective permeability exceeding 10,000 at 1 kHz. In addition, the addition of 2 to 4 %Nb to (Fe,Co)-B-Si alloys was found to cause the change from amorphous to glassy type and the glassy alloy rod of 5 mm in diameter in conjunction with a supercooled liquid region of 50 K was synthesized for Fe-Co-B-Si-Nb alloy [44]. The maximum diameter increases further to 7.7 mm by flux melting, followed by water quenching [45]. The Fe-rich glassy alloys exhibit a rather high B_s above 1.3 T, while the Co-rich glassy alloys have very low coercivity below 1 A/m because of nearly zero saturation magnetostriction (λ_s) of the order 10^{-7}. The Co-rich Co-Fe-B-Si-Nb glassy alloy also exhibits rather good high-frequency effective permeability of about 20,000 at the high frequency of 100 kHz. The Fe-Co-B-Si-Nb glassy alloy thick sheet with a thickness of 0.5 mm and a width of 10 cm was also prepared by twin-roller casting [46]. The glassy alloy ring core made from the thick sheet shows rather good soft magnetic properties of 1.6 A/m for coercivity and 170,000 for maximum permeability in an annealed state, being attractive for soft magnetic thick ring core materials.

In addition, the addition of 4 %Dy to Fe-B-Si-Nb glassy alloy caused a significant increase in saturation magnetostriction to 65×10^{-6} in conjunction with glass-forming ability of 4 mm in diameter and large supercooled liquid region of 100 K [47]. The bulk glassy alloy exhibiting simultaneously high λ_s, GFA, fracture strength and large ΔT_x is attractive for application to sensors.

Here it is important to point out that the Mo-free glassy alloy in the development process of Fe-Mo-(P,C,B,Si) bulk glassy alloys also has a bulk glass-forming ability of 2.5 mm in diameter and high saturation B_s of 1.51 T [48]. Further fine adjustment of alloy component to Fe-rich side causes the formation of Fe-P-B-Si-C glassy alloy with a high saturation magnetization reaching 1.6 T and a maximum diameter of 1 mm [49].

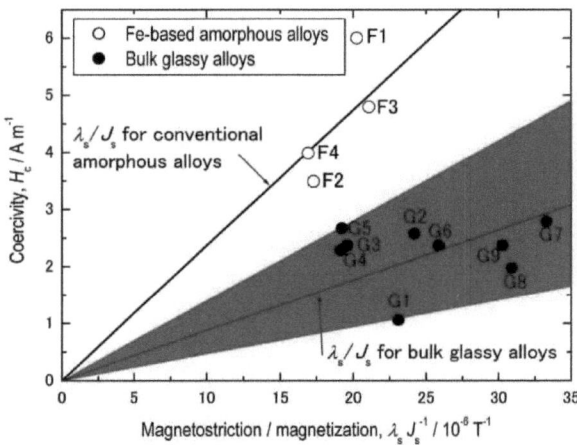

Fig. 7 Comparison of the relations between coercivity and
magnetostriction/magnetization for glassy and amorphous alloys in
Fe-based system [50].

G1	$Fe_{80}P_{12}B_4Si_4$	G2	$Fe_{76}Al_4P_{12}B_4Si_4$
G3	$Fe_{73}Al_5Ga_2P_{11}C_5B_4$	G4	$Fe_{72}Al_5Ga_2P_{11.55}C_{5.2}B_{4.2}$
G5	$Fe_{73}Al_{2.86}Ga_{1.14}P_{12.65}C_{5.75}B_{4.6}$	G6	$Fe_{77}Al_{2.14}Ga_{0.84}P_{8.4}C_5B_4Si_{2.6}$
G7	$Fe_{78}Al_2P_{10}B_6Ge_4$	G8	$Fe_{75}Al_5P_{10}B_6Ge_4$
G9	$Fe_{73}Al_5Ga_2P_{10}B_6Ge_4$	G10	$[(Fe_{0.5}Co_{0.5})_{0.75}B_{0.2}Si_{0.05}]_{96}Nb_4$
G11	$[(Fe_{0.6}Co_{0.4})_{0.75}B_{0.2}Si_{0.05}]_{96}Nb_4$	G12	$[(Fe_{0.7}Co_{0.3})_{0.75}B_{0.2}Si_{0.05}]_{96}Nb_4$
G13	$[(Fe_{0.8}Co_{0.2})_{0.75}B_{0.2}Si_{0.05}]_{96}Nb_4$	G14	$[(Fe_{0.9}Co_{0.1})_{0.75}B_{0.2}Si_{0.05}]_{96}Nb_4$
F1	$Fe_{80}B_{20}$	F2	$Fe_{78}B_{13}Si_9$
F3	$Fe_{80}B_{13}C_7$	F4	$Fe_{80}P_{16}C_3B_1$

In order to clarify the feature of soft magnetic properties of Fe-based
bulk glassy alloys, the relation between coercivity and the ratio of saturated
magnetostriction to saturated magnetization was summarized in Fig. 7 for
various Fe-based bulk glassy alloys, in comparison with conventional
amorphous alloys which need high cooling rates for their formation, on the

basis of coercivity $H_c \propto \Delta V \sqrt{\rho_d} \dfrac{\lambda_s}{J_s}$ [50]. Here, ΔV is volume of defects (internal stress), ρ_d is density of defects, λ_s is saturated magnetostriction and J_s is saturated magnetization.

The slope in the relation is correlated to the volume and density of internal defects. As seen in the figure, the slope is much smaller for the bulk glass type alloys than that for the amorphous type alloys. The distinct difference indicates that the bulk glass type alloys include much lower volume and density of internal defects. It is thus concluded that the formation of more homogeneous and highly dense glassy structure resulting from the alloy components satisfied with the three component rule is the origin for achievement of the good soft magnetic properties with much lower coercivities than those for amorphous type alloys.

8. Structural Fe-based Bulk Glassy Alloys

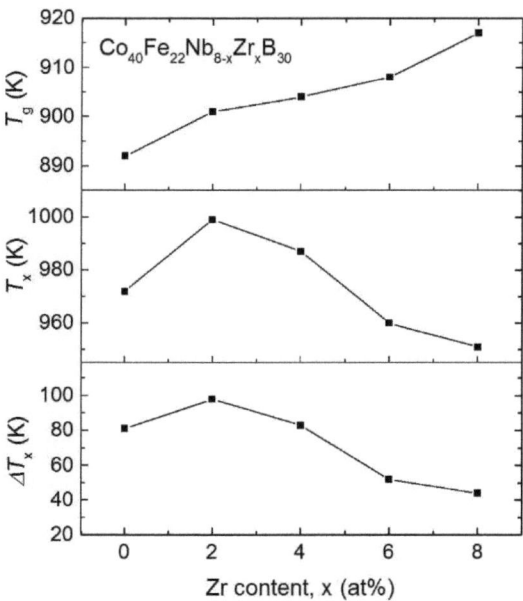

Fig. 8 Changes in T_g, T_x and ΔT_x as a function of Zr content for the melt-spun $Co_{40}Fe_{22}Nb_{8-x}Zr_xB_{30}$ (x=0, 2, 4, 6, and 8 at%) glassy alloys [53].

As required characteristics for structural materials, one can list up high strength at room temperature and elevated temperatures, good wear resistance, high corrosion and oxidation resistances, high glass-forming ability, good superplastic-forming ability and good ductility. For instance, Fe-ETM-B bulk glassy alloys in Fe-Nb-B, Fe-Zr-B and Fe-Nb-Zr-B systems exhibit a rather high T_g of 870 K and large ΔT_x of 91 K, in addition to high hardness [51]. More multiplication of alloy components causes an increase of maximum diameter to 6 mm for Fe-Co-Zr-Mo-W-B system [52]. Co-rich Co-Fe-(Nb,Zr)-B bulk glassy alloys exhibit a high T_g of 900 K, high T_x of 1000 K and large supercooled liquid region of 100 K, as shown in Fig. 8 [53]. These alloys are attractive as structural materials which satisfy the above-described necessary factors.

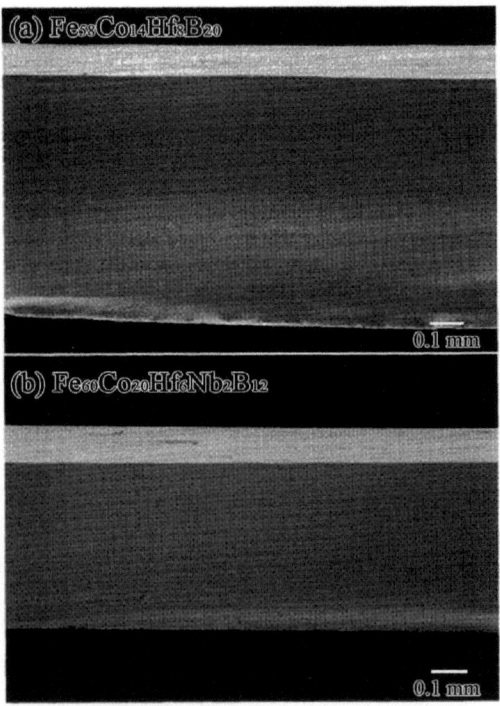

Fig. 9 SEM images of the cross sections of the thick sheets prepared by squeeze casting. (a) $Fe_{58}Co_{14}Hf_8B_{20}$ and (b) $Fe_{60}Co_{20}Hf_6Nb_2B_{12}$ [54].

In addition, Fe-based bulk glassy alloy sheets with a thickness of 0.5 mm in Fe-Co-Hf-B and Fe-Co-Hf-Nb-B systems have been formed by squeeze casting [54]. As seen in Fig. 9, the cast glassy thick sheet has a uniform outer shape and very smooth surface which are suitable for application to molding materials [54].

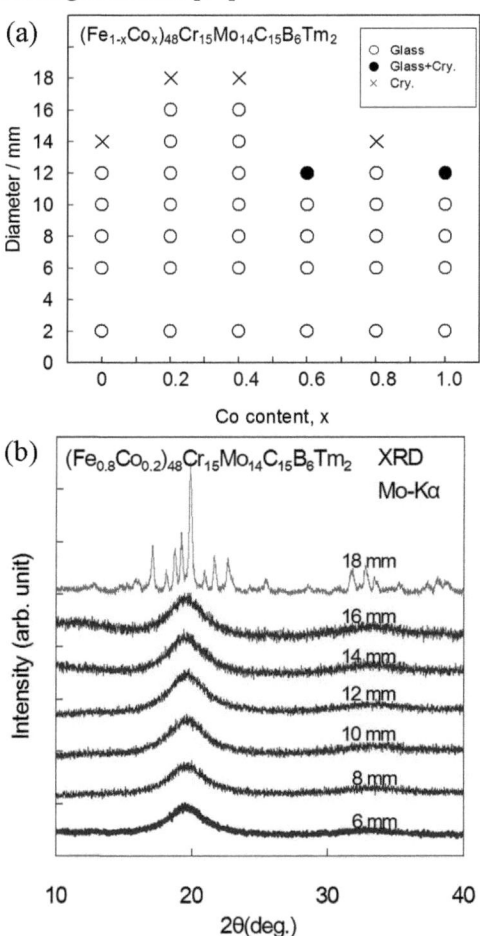

Fig. 10 (a) Change in the structure of the $(Fe_{1-x}Co_x)_{48}Cr_{15}Mo_{14}C_{15}B_6Tm_2$ rods with sample diameter and Co content. (b) X-ray diffraction data of $(Fe_{0.8}Co_{0.2})_{48}Cr_{15}Mo_{14}C_{15}B_6Tm_2$ rods with different diameters [58].

It has also been reported that glassy alloys are formed in Fe-Cr-Mo-C-B system which is attractive for coating materials exhibiting a high corrosion resistance, high hardness and high wear resistance [55, 56]. In $Fe_{43}Cr_{16}Mo_{16}(C,B,P)_{25}$ alloy series, the largest supercooled liquid region exceeding 90 K is obtained around $Fe_{43}Cr_{16}Mo_{16}C_{15}B_{10}$ and the reduced glass transition temperature (T_g/T_l) and the maximum diameter are above 0.6 and 2.7 mm, respectively. The Fe-Cr-Mo-C-B glassy alloy exhibits better corrosion resistance and much higher hardness than those for SUS 304 steel [55].

Later it has been found that the addition of Er or Y to Fe-Cr-Mo-C-B [37] and Fe-Cr-Mo-Mn-C-B [57] glassy alloys causes a significant increase of maximum diameter to over 10 mm. We have also found that Tm element has a more effectiveness on the increase of the maximum diameter as is evidenced from the data shown in Fig. 10 on the formation of bulk glassy alloys with diameters of 10 to 16 mm for $(Fe_{1-x}Co_x)_{48}Cr_{15}Mo_{14}C_{15}B_6$ alloys [58]. The maximum diameter increases further to 18 mm through fine adjustment of alloy component to $(Fe_{0.8}Co_{0.2})_{47}Cr_{15}Mo_{14}C_{15}B_6Tm_3$ [59].

As basic data for structural materials, it seems important to introduce good mechanical properties of glassy Fe-C-Si-P-B alloys with diameters of 0.5 to 0.7 mm after full crystallization [34]. The addition of 1.5 at% B to conventional cast iron FC20 enables the production of amorphous alloy rods of 0.5 to 0.7 mm in diameter and the subsequent annealing for 3.6 ks at 1273 K results in ductile alloy rods consisting of a mixed structure of α-Fe + Fe_3C + graphite. The as-cast glassy alloy rods exhibit a high tensile fracture strength of about 3500 MPa in the absence of appreciable plastic elongation, while the mixed crystalline phase alloy rods exhibit a high yield strength of about 1300 MPa, high ultimate tensile strength of 1530 MPa and large elongation of about 9 %, in addition to a good bending plasticity. It is also expected that their mechanical strength increases significantly by cold drawing treatment. Thus, the Fe-based crystalline alloy rods with good mechanical properties produced by the simple process of casting and annealing are attractive for future application as structural materials.

9. Features of Structure and Crystallization

The formation of bulk glassy alloys in various Fe-based systems is due to the development of unique medium-range ordered atomic configurations like network atomic configurations in which distorted trigonal prisms and anti-Archimedean prisms are connected each other with edge- and face-shared configuration modes [60-62]. Besides, the primary crystallization phase from glassy phase also has a feature, namely, $Fe_{23}B_6$ phase consisting of anti-Archimedean prisms for $Fe_{70}Nb_{10}B_{20}$ [60, 61], $(Fe_{0.5}Co_{0.5})_{72}B_{20}Si_4Nb_4$ [63] and $Fe_{75}Mo_4P_{10}C_4B_4Si_3$ [64] glassy alloys with maximum diameters up to 5 mm. It has further been clarified that the primary phase changes to a chai (χ) phase with a complex cubic structure including icosahedral-like atomic configurations for the Fe-Cr-Mo-C-B-Tm bulk glassy alloys with maximum diameters above 10 mm [65, 66]. Thus, the development of such medium-range ordered atomic configurations caused by choosing suitable alloy components plays an important role in the formation of bulk glassy alloys. In particular, the significant effect of rare earth elements (Er, Y, Tm) has been thought to be attributed to the suppression of crystalline precipitates through the difficulty atomic rearrangements resulting from the dissolution of rare earth elements with very large atomic sizes, in addition to the scavenging effect to impurity elements [65, 66].

10. Applications of Engineering Bulk Glassy Alloys

It was shown that Fe-based bulk glassy alloys exhibit various unique and useful properties such as rather high glass-forming ability, low coercivity, high maximum permeability, high frequency permeability, low core loss, rather high saturated magnetostriction, high static and dynamic fracture strength, large elastic strain, viscous flow deformability and high electrical resistivity, though the saturation magnetization is below about 1.5 T and the fracture toughness is rather low. Based on the understandings of these advantage and disadvantage points, Fe-based bulk glassy alloys have already been commercialized in various fields with trademarks of Liqualloy

[67] and SENNTIX [68] for ferromagnetic alloys and GALOA [69] and AMO-beads [70] for non-ferromagnetic alloys. Typical alloy systems are Fe-(Cr,Nb,Mo)-(P,C,B,Si) for Liqualloy and SENNTIX and Fe-(Cr,Mo)-(C,B) for GALOA and (Fe,Ni)-(Cr,Mo)-(B,Si) for AMO-beads.

The Liqualloy powder consolidated cores made of water-atomized glassy alloy powders and resin have been commercialized in DC/DC convertors in notebook PCs and servers etc. and Point of load type power suppliers in high efficient and low heat type power inductor series. Besides, as high efficient and large current type power inductor series, the Liqualloy cores have been used as CPU and graphic power supply circuits for notebook personal computers, DC/DC convertors for servers and game consoles etc. and point of load type power supplies. These cores have a wide size range of about 5 mm to 12 mm in one edge dimension. The high efficiency is due to the higher values of real (μ') and imaginary (μ'') permeability in the higher frequency range above about 500 kHz as compared with high performance Fe-Si-Al Sendust core, which is shown in Fig. 11 [28, 67].

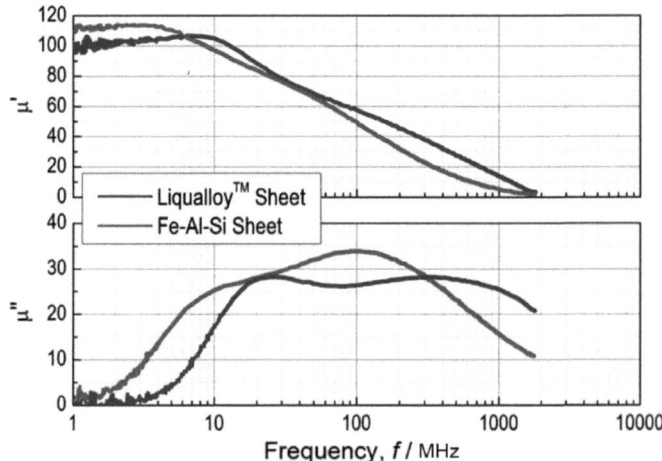

Fig. 11 Higher real part of permeability and higher imaginary part of permeability of "Liqualloy" sheet in the higher frequency range of over 500 MHz as compared with an Fe–Si–Al Sendust sheet. [28]

19

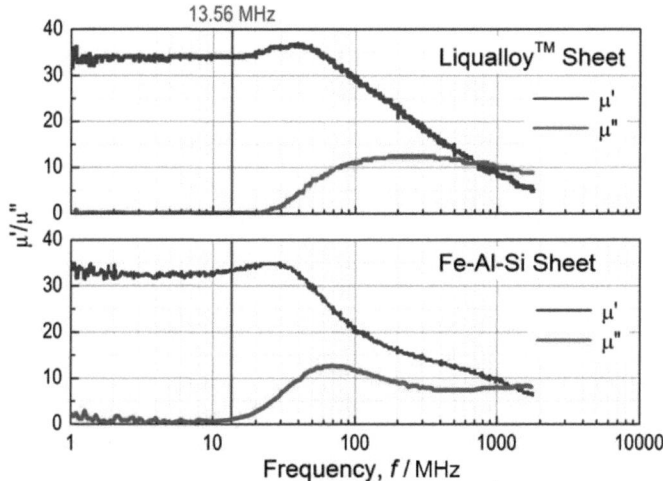

Fig. 12 Good antenna sensitivity results from the much higher quality factor defined by the ratio of real part of permeability to imaginary part of permeability, the data of Sendust sheet are shown for comparison [28].

The thin composite sheet consisting of flaky Liqualloy powder embedded in resins exhibits a high conversion ratio from electro-magnetic noise to heat, indicating a high noise suppression effect. The high conversion ratio comes from the much higher quality factor defined by the ratio of μ'/μ''. The μ'/μ'' value obtained from the data in Fig. 12 is about 100 for Liqualloy sheet and about 30 for the conventional Sendust sheet. Thus, the Fe-based glassy alloy composite sheet exhibits very high magnetic performance in a high frequency field owing to the simultaneous achievements of good magnetic softness, high electrical resistivity and unique powder shape with very thin thickness, large aspect ratio and smooth outer surface etc. Therefore, the Liqualloy sheet has been used as a noise suppression sheet in various electro-magnetic instruments such as digital still camera etc. Besides, the Liqualloy sheet has been used in the radio frequency identification (RFID) system. The insertion of the Liqualloy sheet between metal parts (electro-magnetic devices) and loop antenna enables the penetration of magnetic flux lines into loop antenna, resulting in a significant increase in antenna sensitivity at a carrier frequency of 13.56 MHz. The RFID system has been applied to NTT DoCoMo FOMA type cell phones etc.

In addition, we have developed another high efficient soft magnetic cores of Fe-(Cr,Nb)-(P,C,B,Si) system with trademark of SENNTIX in collaboration with NEC TOKIN corporation. The application of the high efficient choke coil enables the reduction of core loss by more than 50 %, the save of thermal issue at PC board by 10 degree centigrade and the extension of battery life time by about 10 %, and hence the magnetic cores have also been commercialized since 2009.

Very recently, the Liqualloy magnetic cores have been trying to extend its application field to HV/EV reactors because of much lower core losses and higher power supply efficiency exceeding those for 6.5%Si steel core.

11. Features and Applications of Co-based Bulk Glassy Alloys

In addition to the above-described features for Fe-based bulk glassy alloys, Co-based bulk glassy alloys have some additional features of higher yield strength, larger supercooled liquid region, nearly zero saturation magnetostriction and low saturation magnetostriction coefficient exhibiting Giant Magneto-Impedance (GMI) effect. For instance, Co-Fe-Ta-B bulk glassy alloys exhibit very high yield strength above 5300 MPa which is the highest strength among all bulk glassy alloys [71]. The Co-based bulk glassy alloys have rather high glass-forming ability of 5.5 mm for ferromagnetic $(Co_{0.7}Fe_{0.3})_{68}B_{21.9}Si_{5.1}Nb_5$ [72] alloy and 10 mm for non-ferromagnetic $Co_{48}Cr_{15}Mo_{14}C_{15}B_6Tm_2$ alloy [58]. The ferromagnetic type bulk glassy alloys exhibit a saturation magnetization of about 0.5 to 0.9 T, low coercivity of 0.7 to 1.6 A/m and high fracture strength of 4200 to 4450 MPa with distinct plastic strains. Besides, the nearly zero magnetostriction alloys in Co-Fe-Nb-B-Si system exhibit much lower coercivity of 0.2 to 0.4 A/m and higher effective permeability of 21,000 to 47,000 at 1 kHz [73]. When we choose an appropriate alloy composition of $Co_{31}Fe_{31}Nb_6Dy_2B_{30}$ as shown in Fig. 13, the supercooled liquid region reaches as large as 130 K in conjunction with good superplastic forming ability as well as good imprinting formability [74].

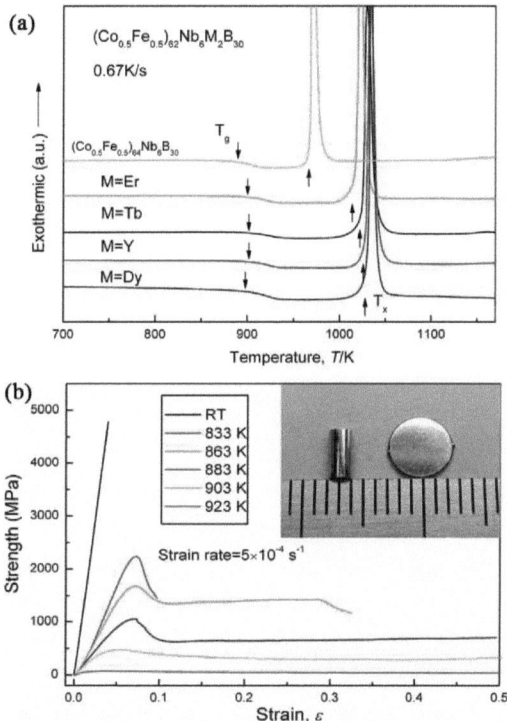

Fig. 13 (a) DSC curves of $Co_{31}Fe_{31}Nb_6M_2B_{30}$ (M = Er, Tb, Y or Dy) and $Co_{31}Fe_{31}Nb_8B_{30}$ glassy alloys. (b) Compressive stress–strain curves of as-cast $Co_{31}Fe_{31}Nb_6Dy_2B_{30}$ rods with a diameter of 2 mm measured at different temperatures. The inset figure shows photographs of the original sample and the sample subjected to compressive deformation at 923 K [74].

The Co-based bulk glassy alloys with nearly zero magnetostriction were tested to apply to various applications such as position sensor, antennas for radio-controlled watch, solenoid valve and magnetic sensor etc. For instance, the magnetic sensor shows a sharper transit signal leading to improved sensitivity and high S/N ratio, higher output voltage leading to improved sensitivity and possibility for miniaturization, and higher strength leading to easy handling and operation. Thus, the magnetic sensor exhibits much better performance characteristics than those for conventional sensors. However, we could not obtain real commercialized states because powder magnetic core forms with lower costs are not suitable for these applications and their materials are much expensive.

12. Applications as Fe-based Structural Materials

AMO-beads have been used as peening shots because the glassy alloy shots exhibit much longer endurance time up to final rupture than those for cast steel shot and high speed steel shot. The glassy alloy shots also have an additional important feature of high protectivity against explosion which is about 8 to 10 times higher than those for cast steel (JIS-Z0311) and high speed steel (JIS-SKH55) shots [75]. By utilizing much longer endurance limit and much higher safety, the AMO-beads have been used as peening shots to high alloy steel gears for racing cars. The application to peening shot treatment causes a significant increase in compressive residual stress to over 2000 MPa as well as an improvement of rotating-beam bending fatigue strength ratio by 50 to 80 %, resulting in a reduction of material weight by about 45 % [28].

In addition, amorphous alloy coated layers of Fe-Cr-Mo-C-B system have been formed on SUS304 or plain carbon steel by the high velocity oxygen fuel method [76]. The glassy alloy coated layer has high relative density exceeding 99 % and a wide thickness range up to about 0.5 mm. The Fe-Cr-Mo-C-B coated glass layer exhibits much better corrosion resistance than that for SUS304 plate and higher Vickers hardness than that for hard chromium plate. In addition, the glassy alloy layer has good adhesion2 with substrate. Neither abrasion nor swelling is detected after cycled bending test under the JIS condition. The higher corrosion resistance, higher hardness and good adhesion characteristics have led to good wear resistance in comparison with hard chromium plating and CrN coating.

Recently, the Fe-based glassy alloy coated layer has also been formed on various nonferrous metals such as aluminum pipe, magnesium plate and polymer etc. In particular, ferromagnetic glassy alloy layers in

Fe-Cr-(P,B,Si) system coated on polymers exhibit good magnetic sensor characteristics. Thus, the coated glassy alloy layers have opened new commercialized fields as corrosion resistant materials, wear resistant materials and sensor materials.

13. Some New Fe-based Alloys

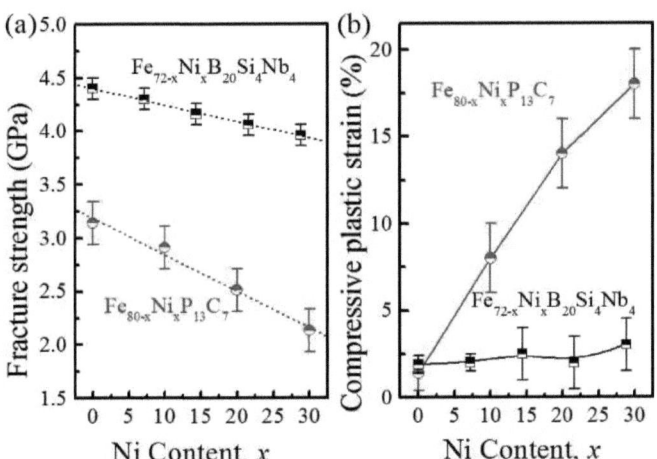

Fig. 14 (a) Change in fracture strength with Ni concentration for $Fe_{80-x}Ni_xP_{13}C_7$ and $Fe_{72-x}Ni_xB_{20}Si_4Nb_4$ BMGs. (b) Change in compressive plasticity with Ni concentration for $Fe_{80-x}Ni_xP_{13}C_7$ and $Fe_{72-x}Ni_xB_{20}Si_4Nb_4$ BMGs [77].

Highly ductile Fe-based bulk glassy alloys have recently been synthesized in Fe-Ni-P-C system. Figure 14 shows the changes in fracture strength and compressive plastic strain with Ni content for $Fe_{80-x}Ni_xP_{13}C_7$ bulk glassy alloys, in comparison with the data of $Fe_{72-x}Ni_xB_{20}Si_{20}Nb_4$ bulk glassy alloys [77]. The fracture strength of the Fe-Ni-P-C bulk glassy alloys decreases from 3140 MPa at 0%Ni to 2130 MPa at 30%Ni, while the plastic strain increases from 1.4% at 0%Ni to about 18% at 30%Ni. This is quite different from the slight increase in plastic strain to 3% for the Fe-Ni-B-Si-Nb alloy containing 30%Ni. The unprecedented plasticity has not been obtained for any other Fe-based bulk glassy alloys. It has been interpreted that the highly ductile nature is due to the shifting from p-d hybridization bonding mode to Ni-Ni and Ni-Fe metallic bonding mode on the basis of XPS and UPS valence-band spectrum data.

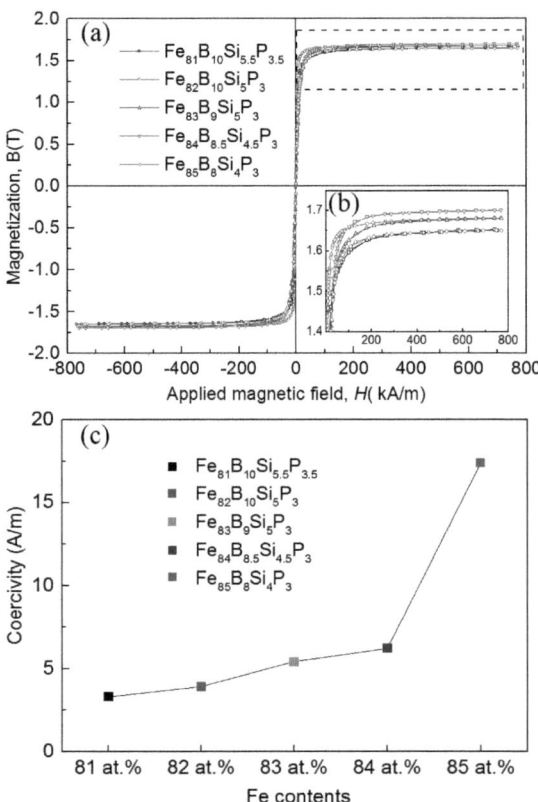

Fig. 15 (a) Hysteresis B-H loops, (b) enlarged partial curves of the B-H loops and (c) coercivity of the ribbons annealed for 600 s at the temperatures of T_x-100 K [78].

In the framework of simple Fe-based amorphous alloys without any transition metals, great efforts have been devoted to develop a new good soft magnetic material exhibiting better engineering performance as compared with conventional soft magnetic amorphous alloys. Figure 15 shows hysteresis B-H loops and coercivity for Fe-rich Fe-metalloid amorphous ribbons produced by melt spinning [78]. The new Fe-based amorphous alloys exhibit simultaneously high saturation magnetization of about 1.7 T and low coercivity below about 5 A/m.

Besides, the amorphous ribbons exhibit good bending ductility even after optimum annealing to achieve good soft magnetic properties. By utilizing these notable characteristics, the application of the new alloys to higher performance magnetic core materials has been trying through the production of wide amorphous alloy sheets with widths of 50 and 65 mm as exemplified in Fig. 16.

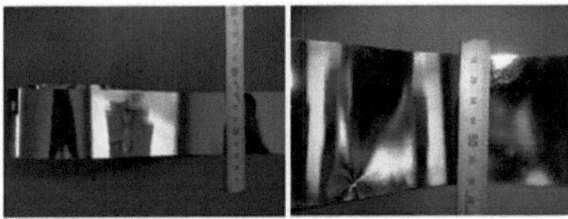

Fig. 16 Photos of amorphous alloy sheets with widths of 50 and 65 mm.

These new amorphous alloy ribbons can be formed in the thickness range up to about 50 μm without appreciable degradation of bending ductility [79]. In addition, the thicker amorphous alloy ribbons exhibit an increase of corrosion resistance in 3.5 mass% NaCl aqueous solution as compared with the amorphous ribbon with a thickness of about 20 μm. The improvement of corrosion resistance caused by the progress of structural relaxation in the absence of any additional transition metals is encouraging for future application of the new soft magnetic amorphous ribbons as higher performance magnetic materials.

14. Fe-based nanocrystalline soft magnetic alloys

Amorphous/glassy alloys are in a metastable state with higher energy and crystallization usually happens to lower their energy under certain conditions such as annealing above the crystallization temperature (T_x) or given sufficient time below T_x. The size and distribution of resulting crystallites are relevant to the crystallization conditions and alloy compositions. In the case of a fixed alloy composition, nanocrystalline alloys with grains in nanoscale embedded in amorphous matrix can be synthesized.

The first Fe-based nanocrystalline alloy reported was Fe-Si-B-Nb-Cu alloy systems in 1988 [80]. The alloy has been attracted great attention because of their unique microstructure and excellent soft magnetic properties such as high permeability, low coercivity (H_c) and core loss. Since then, various nanocrystalline alloys have been developed in Fe-based alloy systems even at present.

This part is intended to summarize developments in synthesis, properties and applications of Fe-based nanocrystalline soft magnetic alloys.

14.1 FINEMET

The Fe-Si-B-Nb-Cu alloy, under a trade name FINEMET®, was known as the first Fe-based soft magnetic nanocrystalline alloy reported. In 1988, Yoshizawa et al. added trace amounts of Cu and M (M=Nb, Mo, W, Ta etc.) with high melting point to the Fe-Si-B alloy, and prepared Fe-Si-B-M-Cu amorphous ribbons by rapid quenching method. Then, the ribbons were annealed in the temperature range from 673 to 923 K for 1 h under a nitrogen gas atmosphere. Cu was added to provide the nuclear site and M could increase the T_x of the amorphous alloy and prevent the overgrowth of the precipitated crystal phase. Among the alloys, the $Fe_{73.5}Si_{13.5}B_9Nb_3Cu_1$ alloy annealed at 853 K showed a microstructure consisting of α-Fe with grain size of about 10 nm distributed in the amorphous matrix. The resulting nanocrystalline alloy exhibited excellent soft magnetic properties, such as low H_c of 0.53 A/m, high μ_e of 100 000 and low core loss of 280 kW/m³.

The crystallization processes of Fe-Si-B-Nb-Cu were studied via various methods, including X-ray diffraction, Mossbauer spectroscopy, small-angle neutron scattering, three-dimensional atom probe (3DAP), transmission electron microscopy (TEM), microprobe analysis and so on [81-84]. A two-stage crystallization is observed.

Amorphous $\rightarrow$ Amorphous$'$ + α-Fe

Amorphous$'$ + α-Fe $\rightarrow$ Fe$_2$B + (Fe,Si)$_3$B + α-Fe

With further increasing the annealing temperature to 1223 K, an FCC Nb-Fe-Si phase was formed. For the first crystallization process, it is generally accepted as Hono's schematic diagram, as is shown in Fig. 17 [83].

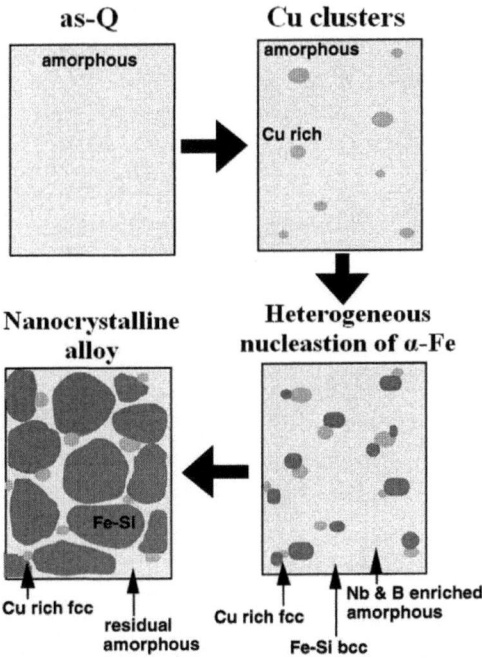

Fig. 17 Schematic diagram of the first crystallization process for Fe-Si-B-Nb-Cu alloy [83].

The effects of alloying addition or substitution in microstructure and properties on FINEMET alloy were intensively studied by many groups

[85-89]. Partial substitution of Co for Fe in Fe-Si-B-Nb-Cu alloy was studied. With 7 at% of Co partial replacement of Fe, the effective hyperfine field at ^{57}Fe nuclei was enhanced and the crystallization process was accelerated. Moreover, the Co-contained ribbon showed low disaccommodation intensity and high initial susceptibility [90]. It is concluded that replacing 7 at% Fe by Co atoms in the classical alloy improves its magnetic properties. Kolano-Burian et al. have studied the nanocrystallization process and soft magnetic properties of $(Fe_{1-x}Co_x)_{73.5}Si_{13.5}B_9Nb_3Cu_1$ alloys. It was found that H_c increased with increasing Co content up to 0.54 and optimal annealing temperature decreases with an increase of Co content [91]. The effect of Co replacement for Fe on the microstructure and soft magnetic properties was also investigated for FINEMET-type alloy with high Fe content by TEM and 3DAP [92]. For $Fe_{78.8-x}Co_xSi_9B_9Nb_{2.6}Cu_{0.6}$ (x=0-60) alloys, B_s values were nearly unchanged, in the range of 1.5-1.55 T, with increasing Co to 40%. After that, it rapidly decreased with further increasing Co content. H_c increased when Co content was higher than 20% and increased drastically above 50%. The changes were considered that the number density of the Cu clusters which serve as heterogeneous nucleation sites decreased with increasing Co content according to the 3DAP analysis [92].

Zbroszczyk et al. have investigated the role of Nb and Cu in the formation of nanocrystalline structure in FINEMET alloy [93]. They concluded that only with the simultaneous addition of Cu and Nb, the alloy with nanocrystalline structure and low H_c can be obtained, and for the resulting nanocrystalline structure, the magnetization vector showed the tendency to align parallel to the ribbon surface, which is required for achieving excellent soft magnetic properties.

The effects of Al addition on the magnetic properties of nanocrystalline Fe-Si-B-Nb-Cu alloy were investigated [94, 95]. The relative initial permeability at 1 kHz increased from 35 000 to 40 000 with 0.1% addition of Al. In the alloy with 2% Al, the H_c decreased to 0.3 A/m and remained in the range of 0.3-0.4 A/m for 2%-10% Al. However, the B_s

values decreased linearly in the 0.1%-10% range. The decrease in coercivity was considered due to the reduction in the intrinsic magneto-crystalline anisotropy K_1, because the effect of Al addition on average grain size is not obvious [94, 95].

Soft magnetic properties of nanocrystalline $Fe_{73.5-x}Mn_xSi_{13.5}B_9Nb_3Cu_1$ were also studied. Partial substitution of Fe by Mn leaded to a decrease in coercivity and an increase in initial permeability and maximum permeability [96]. However, with increasing Mn content, the ribbons become more brittle [97].

The influence of substituting Ge for B and Si on crystallization process and soft magnetic properties were studied [98, 99]. The addition of Ge leaded to a decrease on the first crystalline onset and an increases around 45 K in Curie temperature of the amorphous phase without lowering the B_s of the alloy.

Table II Magnetic properties of FINEMET-type alloys

Alloy (at%)	B_s (T)	H_c (A/m)	μ_e (10^3)	Reference
$Fe_{73.5}Si_{13.5}B_9Nb_3Cu_1$	1.24	0.53	100	[80]
$Fe_{73.5}Si_{16.5}B_6Nb_3Cu_1$	1.18	1.1	75	[80]
$Fe_{78.8}Si_9B_9Nb_{2.6}Cu_{0.6}$	1.52	4	7.4	[92]
$Fe_{68.8}Co_{10}Si_9B_9Nb_{2.6}Cu_{0.6}$	1.52	1.6	/	[92]
$Fe_{58.8}Co_{30}Si_9B_9Nb_{2.6}Cu_{0.6}$	1.53	8	/	[92]
$Fe_{73.4}Si_{13.5}B_9Nb_3Cu_1Al_{0.1}$	1.11	0.9	40	[94]
$Fe_{71.5}Si_{13.5}B_9Nb_3Cu_1Al_2$	1.32	0.3	/	[95]
$Fe_{68.5}Si_{13.5}B_9Nb_3Cu_1Mn_5$	/	1.8	41	[96]
$Fe_{73.5}Si_{13.5}B_9Nb_3Cu_{0.5}Zn_{0.5}$	/	4.8	11	[88]
$Fe_{73.5}Si_{13.5}B_9Nb_3Cu_{0.5}Ag_{0.5}$	/	3.7	25	[88]
$Fe_{73.5}Si_{13.5}B_9Nb_3Au_1$	/	1.8	19	[88]
$Fe_{73.5}Si_{13.5}B_9Nb_{1.5}Cu_1V_{1.5}$	1.26	0.79	135	[101]
$Fe_{72.5}Si_{10}B_{12.5}Nb_4Cu_1$	1.23	0.71	80	[100]
$Fe_{72.5}Si_{10}B_{12.5}Nb_4Cu_1$ *	1.21	1.8	32	[100]

* in bulk form

The possibility of forming a nanocrystalline Fe-based bulk alloy with good soft magnetic properties was examined by Inoue et al [100]. An amorphous alloy rod with a diameter of 0.5 mm was obtained for the $Fe_{72.5}Si_{10}B_{12.5}Nb_4Cu_1$ alloy by copper mold casting. The nanocrystalline alloy consisting of α-Fe and remaining amorphous phases exhibited good soft magnetic properties, such as high B_s of 1.21 T, low H_c of 1.8 A/m and high μ_i of 32000 [100]. The magnetic properties of FINEMET-type alloys developed were summarized in Table II.

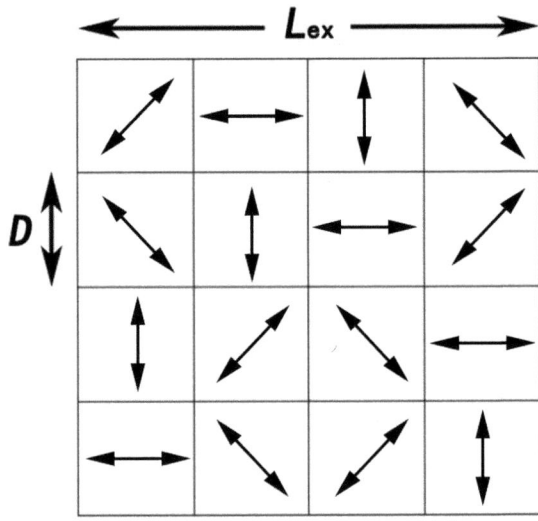

Fig. 18 Schematic representation of the random anisotropy model. The arrows indicate the randomly fluctuating magneto-crystalline anisotropies [104].

The theory why the FINEMET nanocrystalline alloy exhibits good soft magnetic properties was also developed. Based on the microstructure and magnetic results, Herzer has studied the relationship between the grain size D and coercivity H_c as well as initial permeability μ_i within the random anisotropy model [102-104]. For nanocrystalline alloy with grain size far less than ferromagnetic exchange length L_{ex}, the ferromagnetic exchange interaction induces the magnetic moments in magnetic domain to align parallel, instead of following the easy directions of each individual grain.

As a result, the magneto-crystalline anisotropy K_1 was averaged out due to the ferromagnetic exchange interaction, as is shown in Fig. 18. The effective magnetic anisotropy $<K>$ can be expressed as the following formula [103].

$$<K> \approx K_1 \left(\frac{D}{L_{ex}} \right)^{3/2} = \frac{K_1^4 \square D^6}{A^3}$$

Where A is the exchange stiffness, L_{ex} is the ferromagnetic exchange length and D is the grain size, respectively.

As a result, H_c and μ_i can be estimated by:

$$H_c = p_c <K> / (\mu_0 M_s) \approx p_c K_1^4 D^6 / (\mu_0 M_s A^3)$$
$$\mu_i = p_c \mu_0 M_s^2 / <K> \approx p_\mu M_s^2 A^3 / (\mu_0 K_1^4 D^6)$$

According to the formula, coercivity H_c is proportional to the sixth power of grain size D, which is quite correspond with the experiment results. The relationship between grain size and coercivity H_c for various soft magnetic alloys is shown in Fig. 19 [104].

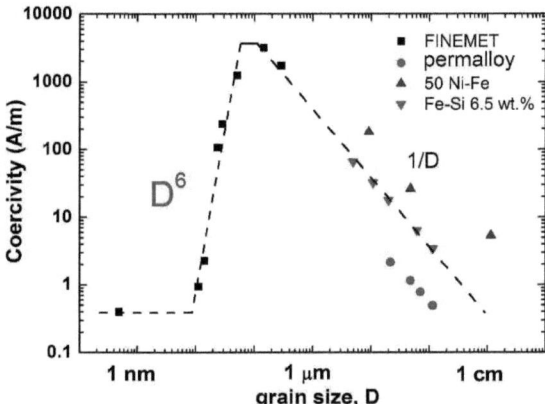

Fig. 19 Grain size and coercivity H_c for various soft magnetic alloys [104].

14.2 NANOPERM

In 1991, nanocrystalline Fe-M-B-Cu (M=Transition Metal, such as Zr,

Hf, Nb, Ta) alloys with nanoscale bcc grain of 10 to 20 nm were prepared by annealing their amorphous precursors [105-107]. The alloys were trade named NANOPERM®. NANOPERM alloys showed excellent soft magnetic properties, such as low H_c of 1-8 A/m, high μ_e of 10 000 to 160 000 and near zero magnetostriction $|\lambda_s|$ of 0.1-1.2×10^{-6}. As compared with FINEMET alloys, NANAOPERM alloys showed a much higher B_s of 1.5 to 1.7 T.

The magnetic properties of NANOPERM alloys are summarized in Table III [105-108].

Table III Magnetic properties of NANOPERM alloys [105-108].

Alloys	B_s (T)	H_c (A/m)	μ_e at 1 kHz
$Fe_{89}Zr_7B_4$	1.65	7.4	15 000
$Fe_{89}Zr_5B_6$	1.70	8.3	13 000
$Fe_{90}Zr_7B_3$	1.63	5.6	22 000
$Fe_{90}Zr_7B_3$*	1.62	26	15 400**
$Fe_{91}Zr_7B_2$	1.70	7.2	14 000
$Fe_{89}Hf_7B_4$	1.59	5.6	32 000
$Fe_{90}Hf_7B_3$	1.59	4.5	30 000
$Fe_{91}Hf_7B_2$	1.60	4.1	18 000
$Fe_{84}Nb_7B_9$	1.49	8.0	22 000
$Fe_{87}Zr_4Nb_3B_6$	1.50	15.9	3 500
$Fe_{86}Zr_7B_6Cu_1$	1.52	3.2	48 000
$Fe_{87}Zr_7B_5Cu_1$	1.55	3.5	20 000
$Fe_{89}Zr_7B_3Cu_1$	1.64	4.5	34 000
$Fe_{90}Zr_7B_2Cu_1$	1.65	2.4	17 000
$Fe_{82}Ti_7B_{10}Cu_1$	1.55	4.9	11 000
$Fe_{83}Nb_7B_9Cu_1$	1.52	3.8	49 000
$Fe_{84}Nb_7B_8Cu_1$	1.56	8.6	13 000
$Fe_{82}Ta_7B_{10}Cu_1$	1.46	8.9	11 000
$Fe_{89}Zr_7B_3Pd_1$	1.63	3.2	30 000
$Fe_{83}Zr_7B_9Ga_1$	1.48	4.8	38 000
$Fe_{88}Zr_7B_3Ni_2$*	1.57	18	22 000**
$Fe_{86}Zr_7B_3Ni_4$*	1.55	14	23 000**
$Fe_{88.7}Zr_7B_3Co_{1.3}$*	1.65	18	14 300**
$Fe_{86}Zr_4Nb_3B_6Cu_1$	1.54	3.7	18 000
$Fe_{86}Zr_{3.25}Nb_{3.25}B_{6.5}Cu_1$	1.61	2.0	110 000
$Fe_{85.6}Zr_{3.3}Nb_{3.3}B_{6.8}Cu_1$	1.57	1.2	160 000
$Fe_{84}Zr_{3.5}Nb_{3.5}B_8Cu_1$	1.53	1.7	120 000

$Fe_{85}Zr_{3.5}Hf_{3.5}B_8Cu_1$	1.44	1.3	92 000
$Fe_{86}Zr_4Nb_3B_6Ni_1$	1.56	5.7	9 300
$Fe_{86}Zr_4Nb_3B_6Pd_1$	1.54	5.1	9 800
$Fe_{86}Zr_4Nb_3B_6Pt_1$	1.47	7.7	10 000
$Fe_{86}Zr_4Nb_3B_6Au_1$	1.51	6.1	11 000
$(Fe_{0.985}Co_{0.015})_{90}Zr_7B_3$	1.64	4.2	27 000
$(Fe_{0.995}Co_{0.005})_{90}Zr_7B_3$	1.62	3.5	34 000
$(Fe_{0.98}Co_{0.02})_{90}Zr_7B_2Cu_1$	1.70	4.5	34 000

* in bulk form, ** μ_{max}

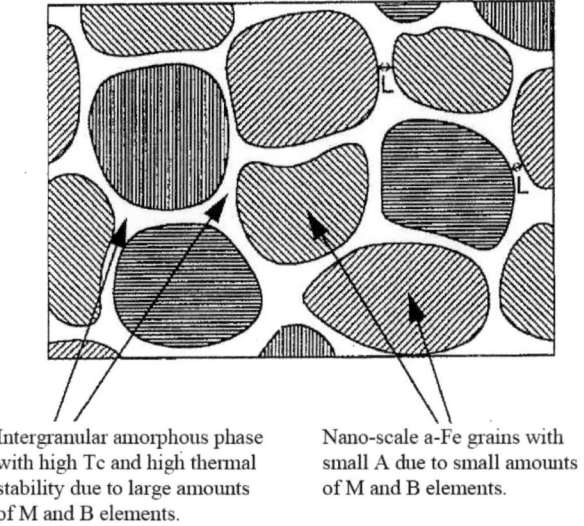

Intergranular amorphous phase
with high Tc and high thermal
stability due to large amounts
of M and B elements.

Nano-scale a-Fe grains with
small A due to small amounts
of M and B elements.

Fig. 20 Schematic diagram of the nanostructure for NANOPERM alloys
[108].

Figure 20 shows the schematic diagram of microstructure for
nanocrystalline NANOPERM alloys [108]. The microstructure is composed
vast majority of α–Fe nanoscale grains and minority of residual
intergranular amorphous phase. The amounts of B and M in the α–Fe grain
phase are much lower than that of the intergranular amorphous phase,
which increases the thermal stability of the residual amorphous phase,
resulting in the suppression of the overgrowth of α–Fe grains. The strong
magnetic exchange coupling between the α–Fe grains through the

intergranular soft magnetic amorphous phase decreases the effective magnetic anisotropy <K> and brings about the good soft magnetic properties. The primary and secondary crystallization reactions of NANOPERM can be expressed as [106]:

$$\text{Amorphous} \rightarrow \text{Amorphous}' + \alpha\text{-Fe}$$
$$\text{Amorphous}' + \alpha\text{-Fe} \rightarrow \text{Fe}_{23}\text{Zr}_6 + \text{Fe}_2\text{Zr} + \alpha\text{-Fe}$$

Figure 21 summarized the expected applications fields for NANOPERM alloys and the magnetic properties which are requested for the applications [108].

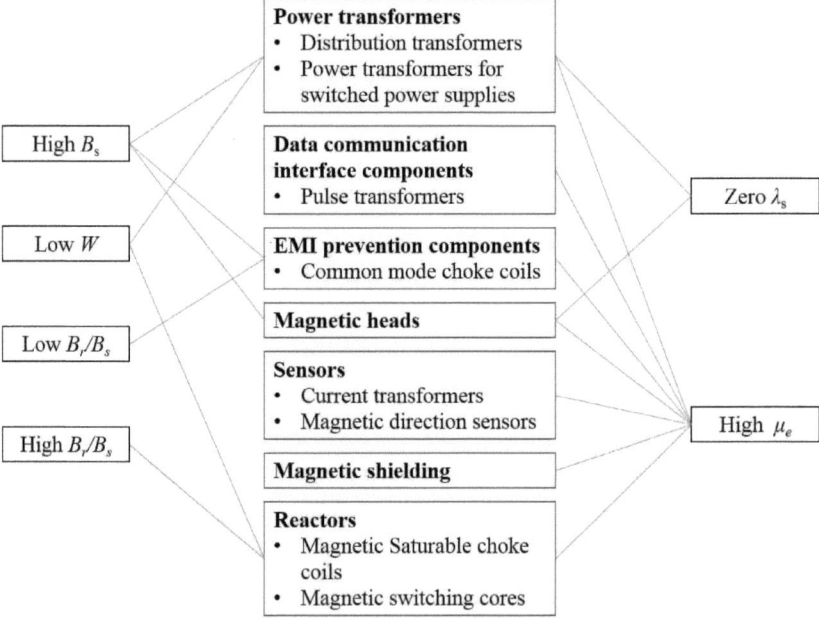

Fig. 21 Application fields for NANOPERM alloys and the magnetic properties which are requested for the applications [108].

14.3 HITPERM

On the basis of NANOPERM alloys, Co was introduced to increase the Curie temperature and B_s, and a new class of nanocrystalline (Fe,Co)–M–B–Cu alloys (M= Zr, Hf, Nb, etc.) have been developed [109, 110]. The alloys were designed for the sake of meeting the demands on soft magnetic materials for high temperature applications and were named HITPERM. The HITPERM alloys exhibit high magnetization until the α→γ occurs at about 1253 K. Compared with commercial HIPERCO-50 alloy (Fe$_{49}$Co$_{49}$V$_2$), HITPERM alloys exhibit good soft magnetic properties in high frequency, such as low H_c of 2.2 Oe at 10 kHz, high ac permeability value of 1800 up to a frequency of 2 kHz and low core of 1 W/g at B_m=1 T and f=10 kHz [110].

The HITPERM alloys also show a two-stage crystallization process. For Fe$_{44}$Co$_{44}$Zr$_7$B$_4$Cu$_1$ alloy, the crystalline process can be expressed as the following sequence [110].

Amorphous $\rightarrow$ Amorphous$'$ + α'-(Fe,Co)

Amorphous$'$+α'-(Fe,Co) $\rightarrow$ (Fe,Co)$_3$Zr+α'-(Fe,Co)

The HITPERM alloys show high thermal stability after long-term annealing. The average grain size only grows up to about 60 nm even in the annealed state for 3072 h at 873 K.

The effects of Fe/Co ratio and field annealing on magnetic properties of HITPERM-type (Fe$_{1-x}$Co$_x$)$_{81}$Nb$_7$B$_{12}$ alloys were investigated [111]. The alloy with Fe/Co ratio close to 1 showed the strongest response to field annealing and induced anisotropies up to 1400 J/m^3 were obtained. The alloys after longitudinal magnetic field annealing exhibited squared hysteresis loops and low H_c in the range 6–10 A/m.

Microstructure of nanocrystalline Fe$_{42.5}$Co$_{42.5}$Nb$_7$B$_8$alloy were investigated by EDS and nano-beam electron diffraction (NBED) with a diameter of 2 nm [112]. Figure 22 shows the EDS results taken from the crystalline region and the amorphous region. It is detected that the amount of Nb in the intergranular amorphous region was larger than that in the crystalline region. As a result, the thermal stability of the amorphous phase increased and the grain growth of the bcc phase was suppressed. The nanocrystalline alloy annealed at 873 K for 3.6 ks consisted of α'-FeCo grain size of 5–10 nm and exhibited a high B_s of 1.9 T as well as a low H_c of 60 A/m. In addition, it showed a very high Curie temperature of above 1173 K.

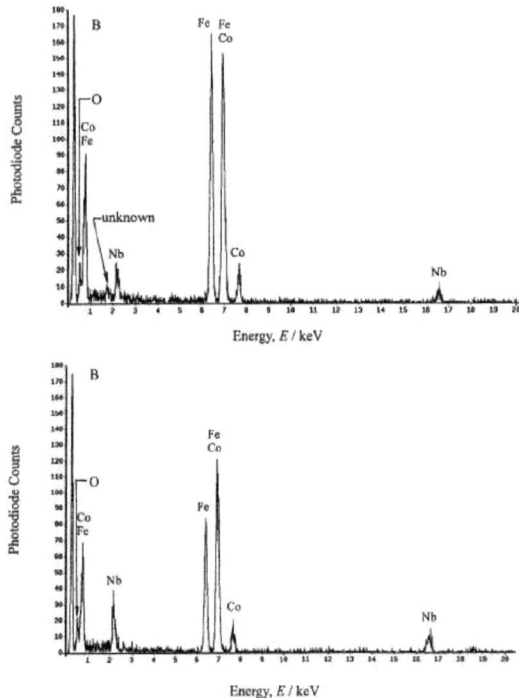

Fig. 22 EDS spectra taken from the bcc phase region and the amorphous phase region [112].

The effect of Ni addition on magnetic properties in HITPERM-type was studied [113]. With substitution of 0.5%-1% Ni for Fe and Co, the soft magnetic properties were significantly improved. After the optimum annealing, H_c decreases to less than 40 A/m and μ_e increases to above 5000. Also, the alloys showed a high B_s of above 1.8 T.

14.4 Newly developed Nanocrystalline alloys

Most recently, several nanocrystalline alloy systems were developed. In this part, the developed Fe-based nanocrystalline soft magnetic alloys were reviewed.

In 2007, by annealing the melt-spinning ribbons of Fe–B–Cu and Fe–Si–B–Cu alloys, nanocrystalline alloys with high B_s of more than 1.8 T and good soft magnetic properties were prepared by Ohta et al [114, 115]. The effect of Cu addition on the microstructure and soft magnetic properties was investigated. With the addition of Cu, the temperature difference between the two crystallization onsets becomes large. As a result, the temperature range to separate out the primary α-Fe nanocrystalline grains becomes wider. With increasing Cu content to 1.5%, the grain size of α-Fe was significantly decreased and the H_c decreases from more than 500 to 7 A/m. Cu plays a very important role in the alloy system.

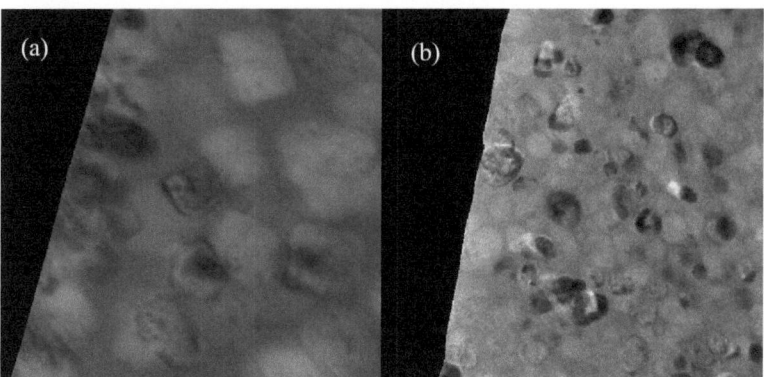

Fig. 23 Cross-section TEM images at surface of $Fe_{82}Si_4B_{12}Cu_1Nb_1$ nanocrystalline alloys: (a) after NA and (b) after HA [117].

The Si/B content ratio dependences of the magnetic properties were studied in Fe–Si–B–Cu alloy [116]. For $Fe_{82.65}Cu_{1.35}Si_xB_{16-x}$ alloy, the temperature difference between T_{x1} and T_{x2} is increased with an increase of Si. In the x=2 alloy, the average grain size of the α-Fe nanocrystalline grains was about 20 nm, and the maximum B_s of 1.84 T and the minimum

H_c of 6.5 A/m were obtained.

The effect of heating rate in the annealing process on microstructure and magnetic properties was examined [117]. The TEM images for $Fe_{82}Si_4B_{12}Cu_1Nb_1$ alloy obtained by normal-heating-rate annealing (NA) and high-heating-rate annealing (HA) are shown in Fig. 23. In the NA alloy, the grain size was as large as 50 nm while that of the HA alloy was about 15 nm. Moreover, the volume fraction of the nanocrystalline phase was larger for the HA alloy. As a result, the HA $Fe_{82}Si_4B_{12}Cu_1Nb_1$ alloy showed a high B_s of 1.78 T and a low H_c 3.2 A/m, whereas the NA alloy was 1.74 T and 120 A/m, respectively. Therefore, the HA method was thought as an effective way to obtain nanocrystalline alloys with good soft magnetic properties.

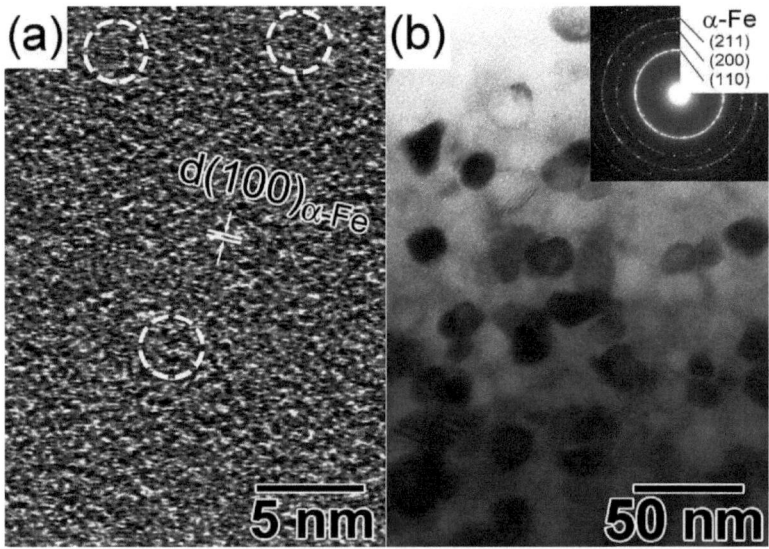

Fig. 24 TEM images of melt-spun $Fe_{83.3}Si_4B_8P_4Cu_{0.7}$ alloy; (a) high resolution image of the as-quenched alloy and (b) bright field image and SAED pattern of the alloy crystallized at 748 K [119].

In late 2007, Makino et al. have investigated the P and Cu addition effect on microstructure and magnetic properties for Fe-Nb-B alloy. It is found that the simultaneous addition of P and Cu significantly decreased the average grain sizes and improved the soft magnetic properties of the annealed alloys [118]. In 2009, P and Cu were added simultaneously to Fe-Si-B alloy with high Fe content, and Fe-Si-B-P-Cu alloys with B_s reached to 1.9 T were prepared [119, 120].

Figure 24 (a) and (b) show the TEM images of the $Fe_{83.3}Si_4B_8P_4Cu_{0.7}$ in the as-quenched and annealed states, respectively. In the melt-spun $Fe_{83.3-84.3}Si_4B_8P_{4-3}Cu_{0.7}$ alloys, heterogeneous amorphous including a large amount of α-Fe clusters of 2–3 nm can be found due to the simultaneous addition of the proper amounts of P and Cu. By annealing the hetero-amorphous alloys, uniform nanocrystalline structure composed of α-Fe grains of about 10 nm could be formed.

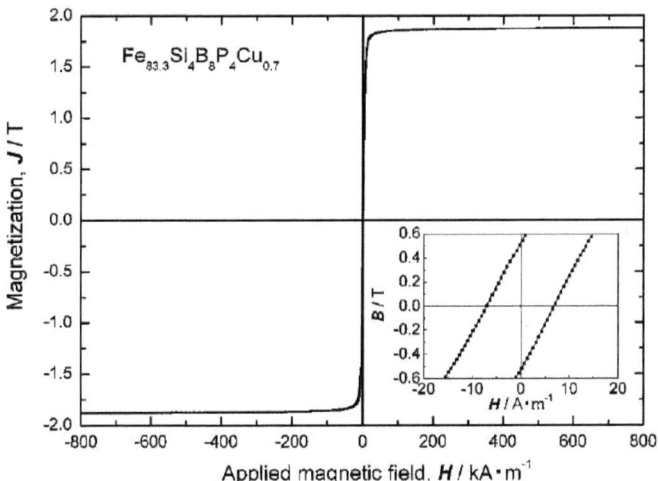

Fig. 25 Hysteresis loop for the nanocrystalline $Fe_{83.3}Si_4B_8P_4Cu_{0.7}$ alloy measured by VSM [121].

The hysteresis loop of the $Fe_{83.3}Si_4B_8P_4Cu_{0.7}$ nanocrystalline alloy is shown in Fig. 25 [121]. The alloy showed a high B_s of 1.88 T due to the high Fe content and H_c of 7 A/m. The B_s is higher than those of the amorphous and the previously reported nanocrystalline alloys. With further increasing Fe content to 84.3%, the B_s of the nanocrystalline alloy reached above 1.9 T, although the H_c increased to 10 A/m.

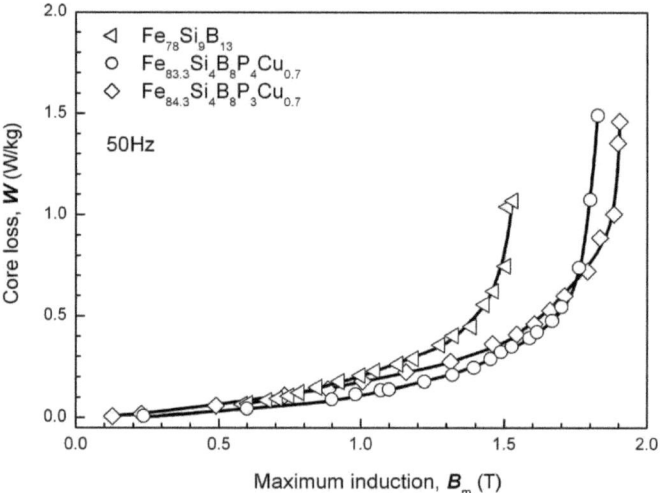

Fig. 26 Core loss of nanocrystalline $Fe_{83.3}Si_4B_8P_4Cu_{0.7}$ alloy as a function of maximum magnetic flux density (B_m). The data of $Fe_{78}Si_9B_{13}$ amorphous alloy annealed at 380 °C for 120 min are also shown for comparison [122].

The core loss was also examined for nanocrystalline $Fe_{83.3}Si_4B_8P_4Cu_{0.7}$ and $Fe_{84.3}Si_4B_8P_3Cu_{0.7}$ alloys [122]. Figure 26 shows the core loss at 50 Hz as a function of maximum magnetic flux density, and the data of $Fe_{78}Si_9B_{13}$ amorphous alloy are shown for comparison. The nanocrystalline alloys exhibit the lower core loss than the amorphous alloy over the whole B_m range examined. The rapid increase of core loss takes place at higher B_m (above 1.7 T) for the nanocrystalline alloy than that (about 1.5 T) for the amorphous $Fe_{78}Si_9B_{13}$ alloy. The outstanding feature of very low core loss in B_m range up to 1.7 T for the nanocrystalline Fe-Si-B-P-Cu alloy should make a contribution to energy saving.

The influences of Cu content and the annealing conditions on magnetic properties and microstructure of Fe-Si-B-P-Cu nanocrystalline alloys were investigated [123]. Figure 27 shows the change in H_c for

$Fe_{84-x}Si_4B_8P_4Cu_x$ alloys annealed for 1.8 ks as a function of annealing temperature (T_A). For the 0.75–1.5 % Cu alloys, the H_c shows a decrease tendency and then increases with further raising temperature, showing a minimum at 723 K.

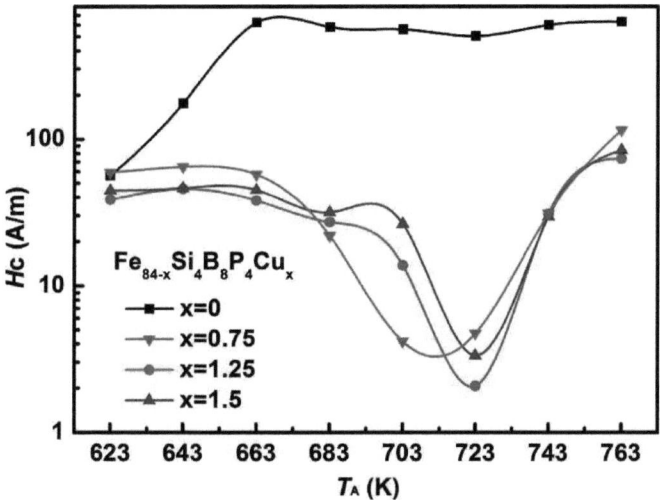

Fig. 27 Changes in Hc for $Fe_{84-x}Si_4B_8P_4Cu_x$ (x=0, 0.75, 1.25, 1.5) alloys as a function of annealing temperature (T_A) for 1.8 ks [123].

The Cu content dependences of (a) B_s, (b) H_c, and (c) average grain size (D) of the α-Fe crystalline phase for the $Fe_{84-x}Si_4B_8P_4Cu_x$ alloys annealed at 723 K for 1.8 ks are shown in Fig. 28, respectively. The values of B_s increase and then decrease with increasing Cu content, and shows a maximum of 1.83 T for the x=1.25 alloy. H_c markedly decreases with an increase of Cu and reaches to a minimum of 2.1 A/m for $Fe_{82.75}Si_4B_8P_4Cu_{1.25}$ alloy. The varying magnetic properties are ascribed to microstructural changes. As shown in Fig. 28 (c), in the Cu-free alloy, the average grain size is above 50 nm. With increasing x, D shows a rapid decrease at x=0.5 and then decreases smoothly. D becomes 23 nm at x=0.5 and less than 20 nm at x=0.75–1.5.

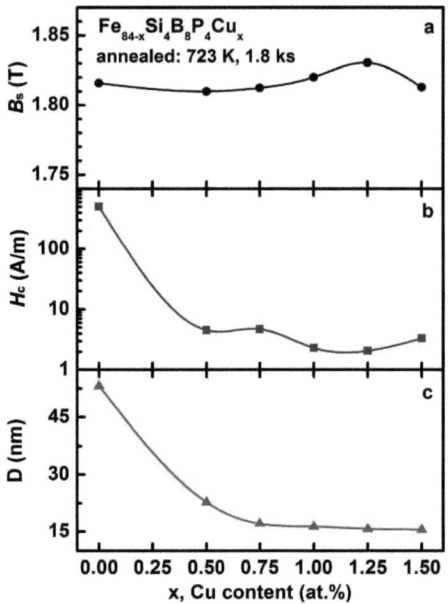

Fig. 28 Cu content dependence of (a) B_s, (b) H_c, and (c) average grain size (D) of α-Fe crystalline phase for the $Fe_{84-x}Si_4B_8P_4Cu_x$ alloys annealed at 723 K for 1.8 ks [123].

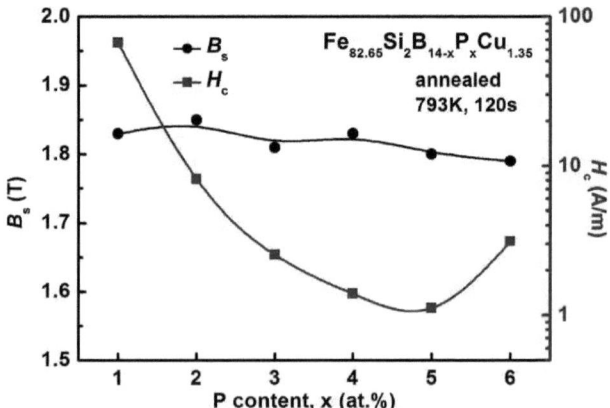

Fig. 29 Changes of B_s and H_c for $Fe_{82.65}Si_2B_{14-x}P_xCu_{1.35}$ alloys annealed at 793 K for 120 s [124].

The effects of P to B concentration ratio on magnetic properties and microstructure in Fe-Si-B-P-Cu nanocrystalline alloys were also investigated [124]. The variations in B_s and H_c of $Fe_{82.65}Si_2B_{14-x}P_xCu_{1.35}$ alloys are shown in Fig. 29. The values of B_s decrease slightly with increasing P content. While the substitution of P for B is effective in decreasing H_c from 67.1 to 1.1 A/m in the range of x=1-5 and increasing to 3.1 A/m with x=6.

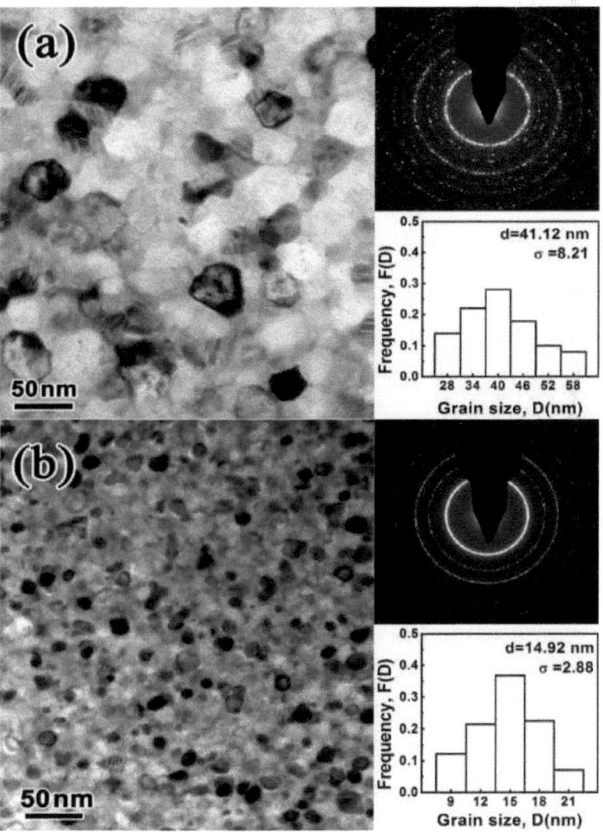

Fig. 30 Bright-field TEM images, selected area electron diffraction (SAED) patterns and distributions of grain sizes of the nanocrystalline (a) $Fe_{82.65}Si_2B_{13}P_1Cu_{1.35}$ and (b) $Fe_{82.65}Si_2B_9P_5Cu_{1.35}$ alloys [124].

Figures 30 (a) and 30 (b) show the bright-field TEM images, selected area electron diffraction (SEAD) patterns and distribution of grain sizes of the nanocrystalline alloys with x=0 and x=5, respectively. In the P free alloy, the average grain size was about 41 nm, whereas, much fine grains of about 15 nm were seen in the alloy with 5% P. The $Fe_{82.65}Si_2B_9P_5Cu_{1.35}$ also showed low core loss of 0.22 W/kg at 50 Hz under B_m=1 T.

Nanocrystalline Fe-B-C-Cu alloys with high magnetic magnetization and good soft magnetic properties have also been developed since 2011 [125]. The Cu contents dependence of (a) H_c and B_s in $Fe_{84-x}B_{10}C_6Cu_x$ alloys are shown in Fig. 31. The B_{800} shows an increasing tendency due to the precipitation of α-Fe, while the H_c decreases with the addition of Cu and exhibits a minimum at x=1.0. The $Fe_{83}B_{10}C_6Cu_1$ nanocrystalline alloy exhibits a high B_s of 1.78 T, low H_c of 5.1 A/m, and low core loss of 3.4 W/kg at 1.0 T and 50 Hz [125, 126].

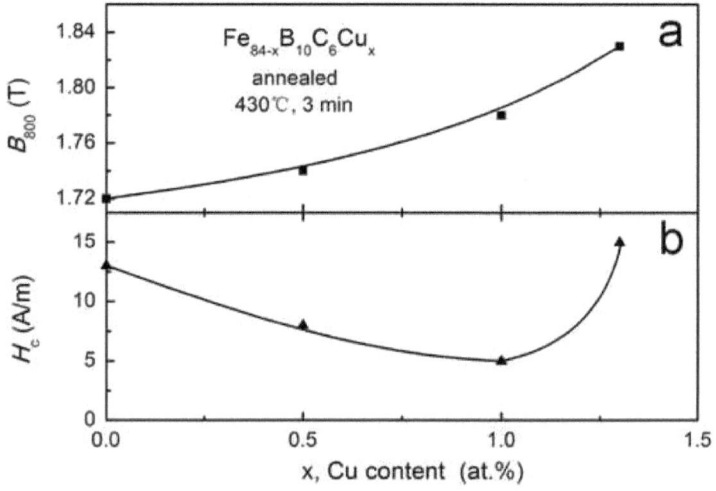

Fig. 31 Dependences of (a) magnetization at 800kA/m (B_{800}) and (b) coercivity (H_c) on Cu content for $Fe_{84-x}B_{10}C_6Cu_x$ alloys annealed at 430 °C for 3min [126].

The influence of Si addition on soft magnetic properties and crystallization behavior for Fe-B-C-Cu alloy was studied [127]. With increasing Si content, H_c decreases slightly and exhibits a minimum value with 2% of Si, meanwhile the μ_e shows a maximum. The B_s shows a slightly decreasing trend owing to the decreasing volume fraction of nanocrystalline phase. The bright-field TEM, HRTEM images and selected area electron diffraction (SAED) pattern of $Fe_{83}B_{10}C_4Si_2Cu_1$ nanocrystalline alloy is shown in Fig. 31. The grain size of α-Fe is about 15 nm. The $Fe_{83}B_{10}C_4Si_2Cu_1$ nanocrystalline alloy exhibits a high B_s of 1.78 T, high μ_e of 13 600, and low H_c of 4 A/m.

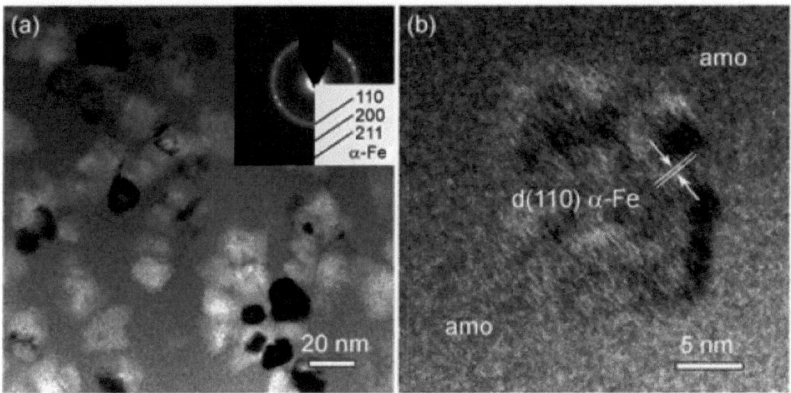

Fig. 31 Bright-field TEM (a), selected area electron diffraction (SAED) pattern and HRTEM images (b) of $Fe_{83}B_{10}C_4Si_2Cu_1$ nanocrystalline alloy [127].

Table IV summarized the B_s, H_c, μ_e and core losses at 1.5 T, 50 Hz and 1.5 T, 400 Hz of the newly developed nanocrystalline alloys. The alloys show high B_s of above 1.7 T, which is higher than that of conventional nanocrystalline alloys. Meanwhile, good soft magnetic properties, such as low H_c and high μ_e can be seen from the table. Although the B_s is a bit lower than Si-steels, the new developed nanocrystalline alloy system shows much superior soft magnetic properties. In addition, the component elements of the alloys are low cost and high productivity. Therefore, the alloys should have a promising application as soft magnetic materials.

Table IV The B_s, H_c, μ_e and core losses of the newly developed nanocrystalline alloys.

Alloy (at%)	B_s (T)	H_c (A/m)	μ_e	$P_{15/50}$ (W/kg)	$P_{15/400}$ (W/kg)	Reference
$(Fe_{0.85}B_{0.15})_{98.5}Cu_{1.5}$	1.83	6.9	>60 000	0.38	/	[114, 128]
$Fe_{82.5}Si_2B_{14}Cu_{1.5}$	1.84	6.5	/	/	/	[115]
$Fe_{80.5}Si_4B_{14}Cu_{1.5}$	1.80	5.7	/	0.26	2	[129]
$Fe_{82.7}Si_2B_{14}Cu_{1.3}$	1.85	6.5	/	0.31	/	[128]
$Fe_{80.6}Si_5B_{13}Cu_{1.4}$	1.8	5.7	/	0.29	/	[128]
$Fe_{82.65}Si_2B_{14}Cu_{1.35}$	1.84	6.5	/	/	/	[116]
$Fe_{82}Si_4B_{12}Nb_1Cu_1$	1.78	3.2	/	0.2	1.3	[130]
$Fe_{83.3}Si_4B_8P_4Cu_{0.7}$	1.88	7	25 000	0.32	/	[131]
$Fe_{84.3}Si_4B_8P_3Cu_{0.7}$	1.94	10	16 000	/	/	[122]
$Fe_{82.75}Si_4B_8P_4Cu_{1.25}$	1.83	2.1	31 600	/	4.6	[123]
$Fe_{83}Si_4B_8P_4Cu_1$	1.82	2.3	30 500	/	4.77	[123]
$Fe_{82.65}Si_2B_9P_5Cu_{1.35}$	1.8	1.1	37 000	0.22	3.26	[124]
$Fe_{83.5}B_{10}C_6Cu_{0.5}$	1.74	8	/	0.37	4.7	[126]
$Fe_{83}B_{10}C_6Cu_1$	1.78	5	/	0.34	4.3	[126]
$Fe_{82.7}B_{10}C_6Cu_{1.3}$	1.83	15	/	0.51	6.9	[126]
$Fe_{83}B_{10}C_4Si_2Cu_1$	1.78	4	13 600	/	/	[127]

15. Conclusion

The present successes and future development of synthesizing Fe- and Co-based amorphous and glassy alloys exhibiting various favorable engineering properties as magnetic and structural materials are expected to contribute to the creation of a low-carbon society.

Acknowledgements

The authors acknowledge technical and financial support from School of Materials Science and Engineering, Tianjin University, China; 1000 Talent Plan for High-Level Foreign Experts, China; Department of Physics, King Abdulaziz University, Saudi Arabia; Highly Cited Scientist Program, King Abdulaziz University, Saudi Arabia; Ningbo Institute of Industrial Technology, Chinese Academy of Sciences, China and Institute for Materials Research, Advanced Institute of Materials research of World Premier Initiative Center, Tohoku University, Japan.

The authors sincerely thank Prof. A.R. Yavari, Prof. A.L. Greer, Prof. D.V. Louzguine, Prof. B.L. Shen, Dr. C.T. Chang and Dr. E. Shalaan for useful discussions and contributions to this review.

References

[1] W. Klement, R.H. Willens, P. Duwez, Non-crystalline Structure in Solidified Gold-Silicon Alloys, Nature, 187 (1960) 869-870.

[2] P. Duwez, S.C.H. Lin, Amorphous Ferromagnetic Phase in Iron-Carbon-Phosphorus Alloys, J. Appl. Phys., 38 (1967) 4096-4097.

[3] S. Hatta, T. Egami, C.D. Graham, Amorphous alloys with improved room-temperature saturation induction, IEEE T. Magn., 14 (1978) 1013-1015.

[4] H. Chen, D. Polk, Novel amorphous metals and amorphous metal articles, United States Patent, Allied Chemical Corporation, 1974, No. US3856513 A.

[5] J.R. Bedell, S. Kavesh, N.S. Hemmat, S. Draizen, R.W. Smith, Contour control for planar flow casting of metal ribbon, United States Patent, Allied Chemical Corporation, 1981, No. US4274473 A.

[6] A. Inoue, J.S. Gook, Fe-based ferromagnetic glassy alloys with wide supercooled liquid region, Mater. Trans., JIM, 36 (1995) 1180-1183.

[7] A. Inoue, M. Hagiwara, T. Masumoto, Production of Fe-P-C amorphous wires by in-rotating-water spinning method and mechanical properties of the wires, J. Mater. Sci., 17 (1982) 580-588.

[8] M. Hagiwara, A. Inoue, T. Masumoto, Mechanical properties of Fe-Si-B amorphous wires produced by in-rotating-water spinning method, Metall. Trans. A, 13 (1982) 373-382.

[9] M. Oguchi, Y. Harakawa, A. Inoue, T. Masumoto, Production of flaky Fe-based amorphous powders by a two-stage quenching technique and their magnetic properties, Mater. Sci. Eng. A-Struct., 181–182 (1994) 1161-1164.

[10] M. Oguchi, A. Inoue, T. Masumoto, Flaky amorphous powders in Fe-, Co-and Al-based systems prepared by a two-stage quenching technique, Mater. Sci. Eng. A-Struct., 133 (1991) 688-691.

[11] A. Inoue, T. Ekimoto, T. Masumoto, Production of Large Spheres by Ejection of Iron-based Liquid Alloy into Circulating Water, T. Iron Steel I. JPN., 27 (1987) 940-945.

[12] A. Inoue, T. Masumoto, S. Arakawa, T. Iwadachi, Amorphous High-Carbon Alloy Steels Rapidly Quenched from Melts, Trans. Jpn. Inst. Met., 19 (1978) 303-304.

[13] M. Naka, K. Hashimoto, A. Inoue, T. Masumoto, Corrosion-resistant amorphous Fe-C alloys containing chromium and/or molybdenum, J. Non-cryst. Solids, 31 (1979) 347-354.

[14] M. Hagiwara, A. Inoue, T. Masumoto, Production of amorphous Co-Si-B and Co-M-Si-B (M=Group IV–VIII transition metals) wires by a method employing melt spinning into rotating water and some properties of the wires, Mater. Sci. Eng., 54 (1982) 197-207.

[15] H.S. Chen, R.C. Sherwood, S. Jin, G.C. Chi, A. Inoue, T. Masumoto, H. Hagiwara, Mechanical properties and magnetic behavior of deformed metal glass wires, J. Appl. Phys., 55 (1984) 1796-1798.

[16] H. Tsubata, S. Tamamura, A. Tanimura, Method and apparatus for continuously manufacturing metal filaments, U.S. Patents, Unitika Ltd. 1986, No. US 4617983 A.

[17] J. Liu, E. Strinning, L. Arnberg, S. Savage, A. Inoue, Production and Properties of Rapidly Solidified Fe-Si Micro-Wires, Scand. J. Metall., 19 (1990) 14-18.

[18] A. Inoue, Y. Kojima, T. Minemura, Formation of Ductile Ni 3 Al Type Compound in Fe-(Ni, Mn)-Al-C Systems by Splat Quenching, Trans. Jpn. Inst. Met., 20 (1979) 468-471.

[19] A. Inoue, H. Tomioka, M. Hagiwara, T. Masumoto, Fabrication and Mechanical Properties of Nonequilibrium Ordered Austenite Wires of Fe-Mn-Al-C System by In-Rotating-Water Spinning Method, Trans. Jpn. Inst. Met., 23 (1982) 341-348.

[20] A. Inoue, H. Tomioka, T. Masumoto, Microstructure and mechanical properties of Fe-Ni-Cr-Al steel wires produced by in-rotating-water spinning method, Metall. Trans. A, 16 (1985) 215-226.

[21] A. Inoue, N. Yano, H. Tomioka, T. Masumoto, Mechanical properties of Fe-Ni-Si-C and Fe-Ni-Cr-Si-C wires prepared by melt spinning, T. Iron Steel I. JPN., 26 (1986) 993-1001.

[22] A. Inoue, M. Oguchi, H. Yamaguchi, T. Masumoto, Aluminum-base amorphous powders with flaky morphology prepared by a two-stage quenching technique, Mater. Trans., JIM, 30 (1989) 1033-1043.

[23] A. Inoue, T. Ekimoto, H.M. Kimura, Y. Masumoto, T. Masumoto, N. Miyake, Preparation of amorphous Ni-Pd-P particles by melt ejection into stirred water and their hardness and thermal stability, Metall. Trans. A, 18 (1987) 377-383.

[24] Y. Li, W. Dong, Y.F. Fu, Y. Tan, A. Miura, A. Kawasaki, The critical cooling rate of Fe-based mono-sized spherical particles with fully glassy phase, Adv. Mat. Res., 509 (2012) 185-191.

[25] Y. Harakawa, A. Inoue, T. Masumoto, Powder-Forming Tendency of Nonequilibrium Phases in Fe-C-X (X= Cr, Mo, W, Al or Si) Alloys by Comminution and Microstructure and Hardness of Their Sintered Compacts, Sci. Rep. RITU, 32 (1985) 277-296.

[26] A. Inoue, T. Iwadachi, T. Minemura, T. Masumoto, Nonequilibrium Phases in Fe-X-C (X=Cr, Mo or W) Ternary Alloys Quenched Rapidly from Melts, Trans. Jpn. Inst. Met., 22 (1981) 192-209.

[27] A. Inoue, L. Arnberg, M. Oguchi, U. Backmark, N. Bäckström, T. Masumoto, Preparation of Fe-Cr-Mo-C amorphous powders and microstructure and mechanical properties of their hot-pressed products, Mater. Sci. Eng., 95 (1987) 101-114.

[28] A. Inoue, A. Takeuchi, Recent development and application products of bulk glassy alloys, Acta Mater., 59 (2011) 2243-2267.

[29] C. Suryanarayana, A. Inoue, Bulk Metallic Glasses, CRC Press, 2011.

[30] A. Inoue, J.S. Gook, Effect of Additional Elements (M) on the Thermal Stability of Supercooled Liquid in $Fe_{72-x}Al_5Ga_2P_{11}C_6B_4M_x$ Glassy Alloys, Mater. Trans., JIM, 37 (1996) 32-38.

[31] A. Inoue, T. Zhang, T. Itoi, A. Takeuchi, New Fe-Co-Ni-Zr-B amorphous alloys with wide supercooled liquid regions and good soft magnetic properties, Mater. Trans., JIM, 38 (1997) 359-362.

[32] T.D. Shen, R.B. Schwarz, Bulk ferromagnetic glasses prepared by flux melting and water quenching, Appl. Phys. Lett., 75 (1999) 49-51.

[33] A. Inoue, W. Zhang, New Fe-based amorphous alloys with large magnetostriction and wide supercooled liquid region before crystallization, J. Appl. Phys., 85 (1999) 4491-4493.

[34] A. Inoue, X.M. Wang, Bulk amorphous FC20 (Fe–C–Si) alloys with small amounts of B and their crystallized structure and mechanical properties, Acta Mater., 48 (2000) 1383-1395.

[35] S. Pang, T. Zhang, K. Asami, A. Inoue, New Fe-Cr-Mo-(Nb, Ta)-CB glassy alloys with high glass-forming ability and good corrosion resistance, Mater. Trans.-JIM, 42 (2001) 376-379.

[36] A. Inoue, B. Shen, Soft magnetic bulk glassy Fe-B-Si-Nb alloys with high saturation magnetization above 1.5 T, Mater Trans, 43 (2002) 766-769.

[37] V. Ponnambalam, S.J. Poon, G.J. Shiflet, Fe-based bulk metallic glasses with diameter thickness larger than one centimeter, J Mater Res, 19 (2004) 1320-1323.

[38] V. Ponnambalam, S.J. Poon, G.J. Shiflet, Fe–Mn–Cr–Mo–(Y, Ln)–C–B (Ln= Lanthanides) bulk metallic glasses as formable amorphous steel alloys, J Mater Res, 19 (2004) 3046-3052.

[39] A. Inoue, High Strength Bulk Amorphous Alloys with Low Critical Cooling Rates, Mater T Jim, 36 (1995) 866-875.

[40] A. Inoue, Stabilization of metallic supercooled liquid and bulk amorphous alloys, Acta Mater., 48 (2000) 279-306.

[41] B. Shen, A. Inoue, Bulk glassy Fe-Ga-PCB-Si alloys with high glass-forming ability, high saturation magnetization and good soft magnetic properties, Mater. Trans., 43 (2002) 1235-1239.

[42] F. Liu, S. Pang, R. Li, T. Zhang, Ductile Fe–Mo–P–C–B–Si bulk metallic glasses with high saturation magnetization, J. Alloy. Compd., 483 (2009) 613-615.

[43] F. Li, B. Shen, A. Makino, A. Inoue, Excellent soft-magnetic properties of (Fe,Co)–Mo–(P,C,B,Si) bulk glassy alloys with ductile deformation behavior, Appl. Phys. Lett., 91 (2007) 234101.

[44] A. Inoue, B.L. Shen, C.T. Chang, Super-high strength of over 4000

MPa for Fe-based bulk glassy alloys in $[(Fe_{1-x}Co_x)_{0.75}B_{0.2}Si_{0.05}]_{96}Nb_4$ system, Acta Mater., 52 (2004) 4093-4099.

[45] T. Bitoh, A. Makino, A. Inoue, A.L. Greer, Large bulk soft magnetic $[(Fe_{0.5}Co_{0.5})_{0.75}B_{0.20}Si_{0.05}]_{96}Nb_4$ glassy alloy prepared by B_2O_3 flux melting and water quenching, Appl. Phys. Lett., 88 (2006) 182510.

[46] A. Urata, N. Nishiyama, K. Amiya, A. Inoue, Continuous casting of thick Fe-base glassy plates by twin-roller melt-spinning, Mater. Sci. Eng. A-Struct., 449–451 (2007) 269-272.

[47] J. Li, H. Men, B. Shen, Soft-ferromagnetic bulk glassy alloys with large magnetostriction and high glass-forming ability, AIP Advances, 1 (2011) 042110.

[48] A. Makino, T. Kubota, C. Chang, M. Makabe, A. Inoue, FeSiBP bulk metallic glasses with high magnetization and excellent magnetic softness, J. Magn. Magn. Mater., 320 (2008) 2499-2503.

[49] J. Zhang, C. Chang, A. Wang, B. Shen, Development of quaternary Fe-based bulk metallic glasses with high saturation magnetization above 1.6 T, J. Non-cryst. Solids, 358 (2012) 1443-1446.

[50] T. Bitoh, A. Makino, A. Inoue, Origin of low coercivity of Fe-(Al, Ga)-(P, C, B, Si, Ge) bulk glassy alloys, Mater. Trans., 44 (2003) 2020-2024.

[51] L. Ma, L. Wang, T. Zhang, A. Inoue, Fe-based metallic glass with significant supercooled liquid region of over 90 K, J. Mater. Sci. Lett, 17 (1998) 1893-1895.

[52] A. Inoue, T. Zhang, H. Koshiba, A. Makino, New bulk amorphous Fe–(Co,Ni)–M–B (M=Zr,Hf,Nb,Ta,Mo,W) alloys with good soft magnetic properties, J. Appl. Phys., 83 (1998) 6326-6328.

[53] T. Itoi, A. Inoue, Thermal stability and soft magnetic properties of Co-Fe-M-B (M=Nb, Zr) amorphous alloys with large supercooled liquid region, Mater. Trans., JIM, 41 (2000) 1256-1262.

[54] H. Fukumura, A. Inoue, H. Koshiba, T. Mizushima, (Fe, Co)-(Hf, Nb)-B glassy thick sheet alloys prepared by a melt clamp forging method, Mater. Trans., 42 (2001) 1820-1822.

[55] K. Asami, S.-J. Pang, T. Zhang, A. Inoue, Preparation and corrosion resistance of Fe-Cr-Mo-C-B-P bulk glassy alloys, J. Electrochem. Soc., 149 (2002) B366-B369.

[56] S.J. Pang, T. Zhang, K. Asami, A. Inoue, Synthesis of Fe–Cr–Mo–C–B–P bulk metallic glasses with high corrosion resistance, Acta Mater., 50 (2002) 489-497.

[57] Z.P. Lu, C.T. Liu, J.R. Thompson, W.D. Porter, Structural Amorphous Steels, Phys. Rev. Lett., 92 (2004) 245503.

[58] K. Amiya, A. Inoue, Fe-(Cr, Mo)-(C, B)-Tm bulk metallic glasses with high strength and high glass-forming ability, Rev. Adv. Mater. Sci, 18 (2008) 27-29.

[59] C. Suryanarayana, A. Inoue, Iron-based bulk metallic glasses, Int. Mater. Rev., 58 (2013) 131-166.

[60] E. Matsubara, S. Sato, M. Imafuku, T. Nakamura, H. Koshiba, A. Inoue, Y. Waseda, Structural study of Amorphous $Fe_{70}M_{10}B_{20}$ (M=Zr, Nb and Cr) alloys by X-ray diffraction, Mater. Sci. Eng. A-Struct., 312 (2001) 136-144.

[61] T. Hanada, Y. Hirotsu, T. Ohkubo, Nanoscale Phase Separation in an Fe70Nb10B20 Glass Studied by Advanced Electron Microscopy Techniques, Mater. Trans., 45 (2004) 1194-1198.

[62] A.R. Yavari, Materials science: A new order for metallic glasses, Nature, 439 (2006) 405-406.

[63] A. Hirata, Y. Hirotsu, K. Amiya, N. Nishiyama, A. Inoue, Change of nanostructure in $(Fe_{0.5}Co_{0.5})_{72}B_{20}Si_4Nb_4$ metallic glass on annealing, Mater. Sci. Forum, 539 (2007) 2077-2081.

[64] B. Shen, M. Akiba, A. Inoue, Excellent soft-ferromagnetic bulk glassy alloys with high saturation magnetization, Appl. Phys. Lett., 88 (2006) 131907.

[65] A. Hirata, Y. Hirotsu, K. Amiya, N. Nishiyama, A. Inoue, Fe23B6-type quasicrystal-like structures without icosahedral atomic arrangement in an Fe-based metallic glass, Phys. Rev. B, 80 (2009) 140201.

[66] A. Hirata, Y. Hirotsu, K. Amiya, A. Inoue, Quasicrystal-like structure

and its crystalline approximant in an $Fe_{48}Cr_{15}Mo_{14}C_{15}B_6Tm_2$ bulk metallic glass, J. Alloy. Compd., 504, S1 (2010) S186-S189.

[67] H. Koshiba;, Y. Naito;, T. Mizushima;, A. Inoue;, Development of the Fe-based Glassy Alloy "LiqualloyTM" and Its Application to Powder Core, Materia Japan, 47 (2008) 39-41.

[68] H. Matsumoto, A. Urata, Y. Yamada, A. Inoue, Novel FePBNbCr glassy alloys "SENNTIX" with good soft-magnetic properties for high efficiency commercial inductor cores, J. Alloy. Compd., 509, S1 (2011) S193-S196.

[69] A. Inoue, Y. Motoe, K. Nakashima, T. Ishikawa, M. Sugiyama, T. Igarashi, H. Wakiwaka, A. Makino, Magnetostrictive film, magnetostrictive element, torque sensor, force sensor, pressure sensor, and process for production of magnetostrictive film, European Patent, 2012. No. EP 2466662 A1.

[70] K. Okumura, J. Kajita, J. Kurosaki, H. Kimura, A. Inoue, Development of Fe-based metallic glass shot "AMO-beads"for peening with high strength and long life, in: 10th Int. conf. on shot peening (ICSP 10th), Tokyo, Japan, 2008.

[71] A. Inoue, B. Shen, H. Koshiba, H. Kato, A.R. Yavari, Cobalt-based bulk glassy alloy with ultrahigh strength and soft magnetic properties, Nat. Mater., 2 (2003) 661-663.

[72] Y. Dong, A. Wang, Q. Man, B. Shen, $(Co_{1-x}Fe_x)_{68}B_{21.9}Si_{5.1}Nb_5$ bulk glassy alloys with high glass-forming ability, excellent soft-magnetic properties and superhigh fracture strength, Intermetallics, 23 (2012) 63-67.

[73] H. Sun, Q. Man, Y. Dong, B. Shen, H. Kimura, A. Makino, A. Inoue, Effect of Nb addition on the glass-forming ability, mechanical and soft-magnetic properties in $(Co_{0.942}Fe_{0.058})_{72-x}Nb_xB_{22.4}Si_{5.6}$ bulk glassy alloys, J. Alloy. Compd., 504, S1 (2010) S31-S33.

[74] Q. Man, A. Inoue, Y. Dong, J. Qiang, C. Zhao, B. Shen, A new CoFe-based bulk metallic glasses with high thermoplastic forming ability, Scripta Mater., 69 (2013) 553-556.

[75] A. Inoue, I. Yoshii, H. Kimura, K. Okumura, J. Kurosaki, Enhanced

shot peening effect for steels by using Fe-based glassy alloy shots, Mater. Trans., 44 (2003) 2391-2395.

[76] S.J. Pang, T. Zhang, K. Asami, A. Inoue, Bulk glassy Fe–Cr–Mo–C–B alloys with high corrosion resistance, Corros. Sci., 44 (2002) 1847-1856.

[77] W. Yang, H. Liu, Y. Zhao, A. Inoue, K. Jiang, J. Huo, H. Ling, Q. Li, B. Shen, Mechanical properties and structural features of novel Fe-based bulk metallic glasses with unprecedented plasticity, Sci. Rep., 4 (2014) 1-6.

[78] F.L. Kong, C.T. Chang, A. Inoue, E. Shalaan, F. Al-Marzouki, Fe-based amorphous soft magnetic alloys with high saturation magnetization and good bending ductility, J. Alloy. Compd., 615 (2014) 163-166.

[79] Y. Han, F.L. Kong, C.T. Chang, S.L. Zhu, A. Inoue, E.S. Shalaan, F. Al-Marzouki, Syntheses and corrosion behaviors of Fe-based amorphous soft magnetic alloys with high saturation magnetization near 1.7 T, J Mater. Res., (2014) (In press).

[80] Y. Yoshizawa, S. Oguma, K. Yamauchi, New Fe-based soft magnetic alloys composed of ultrafine grain structure, J. Appl. Phys., 64 (1988) 6044-6046.

[81] T.H. Noh, M.B. Lee, H.J. Kim, I.K. Kang, Relationship between crystallization process and magnetic properties of Fe‐(Cu‐Nb)‐Si‐B amorphous alloys, J. Appl. Phys., 67 (1990) 5568-5570.

[82] G. Rixecker, P. Schaaf, U. Gonser, Crystallization behaviour of amorphous $Fe_{73.5}Cu_1Nb_3Si_{13.5}B_9$, J Phys.-Condens. Mat., 4 (1992) 10295.

[83] K. Hono, D.H. Ping, M. Ohnuma, H. Onodera, Cu clustering and Si partitioning in the early crystallization stage of an $Fe_{73.5}Si_{13.5}B_9Nb_3Cu_1$ amorphous alloy, Acta Mater., 47 (1999) 997-1006.

[84] M. Ohnuma, K. Hono, S. Linderoth, J.S. Pedersen, Y. Yoshizawa, H. Onodera, Small-angle neutron scattering and differential scanning calorimetry studies on the copper clustering stage of Fe-Si-B-Nb-Cu nanocrystalline alloys, Acta Mater., 48 (2000) 4783-4790.

[85] V. Franco, C.F. Conde, A. Conde, Magnetic properties and nanocrystallization of a $Fe_{63.5}Cr_{10}Si_{13.5}B_9Cu_1Nb_3$ alloy, J. Magn. Magn.

Mater., 203 (1999) 60-62.

[86] J.M. Borrego, A. Conde, I. Todd, M. Frost, H.A. Davies, M.R.J. Gibbs, J.S. Garitaonandia, J.M. Barandiaran, J.M. Greneche, Nanocrystallite compositions for Al- and Mo-containing Finemet-type alloys, J. Non-cryst. Solids, 287 (2001) 125-129.

[87] J.S. Blazquez, J.M. Borrego, C.F. Conde, A. Conde, J.M. Greneche, On the effects of partial substitution of Co for Fe in FINEMET and Nb-containing HITPERM alloys, J. Phys.-condens. Mat., 15 (2003) 3957-3968.

[88] N. Chau, N.Q. Hoa, N.D. The, L.V. Vu, The effect of Zn, Ag and Au substitution for Cu in Finemet on the crystallization and magnetic properties, J. Magn. Magn. Mater., 303 (2006) E415-E418.

[89] R. Brzozowski, M. Wasiak, H. Piekarski, P. Sovak, P. Uznanski, M.E. Moneta, Properties of Mn-doped FINEMET, J. Alloy. Compd., 470 (2009) 5-11.

[90] J. Zbroszczyk, K. Narita, J. Olszewski, W. Ciurzyńska, W. Lijun, B. Wysłocki, S. Szymura, M. Hasiak, Effect of Co addition on the microstructure and magnetic properties of Fe-Cu-Nb-Si-B alloy, J. Magn. Magn. Mater., 160 (1996) 281-283.

[91] A. Kolano-Burian, T. Kulik, G. Vlasak, J. Ferenc, L.K. Varga, Effect of Co addition on nanocrystallization and soft magnetic properties of $(Fe_{1-x}Co_x)_{73.5}Cu_1Nb_3Si_{13.5}B_9$ alloys, J. Magn. Magn. Mater., 272–276, Part 2 (2004) 1447-1448.

[92] M. Ohnuma, D.H. Ping, T. Abe, H. Onodera, K. Hono, Y. Yoshizawa, Optimization of the microstructure and properties of Co-substituted Fe–Si–B–Nb–Cu nanocrystalline soft magnetic alloys, J. Appl. Phys., 93 (2003) 9186-9194.

[93] J. Zbroszczyk, H. Fukunaga, J. Olszewski, W.H. Ciurzyńska, M. Hasiak, The role of Nb and Cu in the creation of nanostructure in Fe-based amorphous alloys, J. Magn. Magn. Mater., 160 (1996) 277-278.

[94] S.H. Lim, W.K. Pi, T.H. Noh, H.J. Kim, I.K. Kang, Effects of Al on the magnetic properties of nanocrystalline $Fe_{73.5}Cu_1Nb_3Si_{13.5}B_9$ alloys, J.

Appl. Phys., 73 (1993) 6591-6593.

[95] B.J. Tate, B.S. Parmar, I. Todd, H.A. Davies, M.R.J. Gibbs, R.V. Major, Soft magnetic properties and structures of nanocrystalline Fe–Al–Si–B–Cu–Nb alloy ribbons, J. Appl. Phys., 83 (1998) 6335-6337.

[96] Anh-Tuan Le, Chong-Oh Kim, Nguyen Chau, Nguyen Due Tho, Nguyen Quang Hoa, H. Lee, Microstructure and Magnetic Characteristics of Mn-doped Finemet Nanocomposites, Journal of Magnetics, 11 (2006) 30-35.

[97] M.R. Tamoria, E.E. Carpenter, M.M. Miller, J.H. Claassen, B.N. Das, R.M. Stroud, L.K. Kurihara, R.K. Everett, M.A. Willard, A.C. Hsiao, M.E. McHenry, V.G. Harris, Magnetism, structure and the effects of thermal aging on $(Fe_{1-x}Mn_x)_{73.5}Si_{13.5}B_9Nb_3Cu_1$ alloys, IEEE T. Magn., 37 (2001) 2264-2267.

[98] V. Cremaschi, G. Sánchez, H. Sirkin, Magnetic properties and structural evolution of FINEMET alloys with Ge addition, Physica B, 354 (2004) 213-216.

[99] D. Muraca, V.J. Cremaschi, H. Sirkin, Effect of the addition of Ge to the FINEMET alloy, J. Magn. Magn. Mater., 311 (2007) 618-622.

[100] A. Inoue, B. Shen, T. Ohsuna, Soft Magnetic Properties of Nanocrystalline Fe-Si-B-Nb-Cu Rod Alloys Obtained by Crystallization of Cast Amorphous Phase, Mater. Trans., 43 (2002) 2337-2341.

[101] W. Lu, B. Yan, Y. Li, R. Tang, Structure and soft magnetic properties of V-doped Finemet-type alloys, J. Alloy. Compd., 454 (2008) L10-L13.

[102] R. Alben, J.J. Becker, M.C. Chi, Random anisotropy in amorphous ferromagnets, J. Appl. Phys., 49 (1978) 1653.

[103] G. Herzer, Grain-Structure and Magnetism of Nanocrystalline Ferromagnets, IEEE T. Magn., 25 (1989) 3327-3329.

[104] G. Herzer, Grain size dependence of coercivity and permeability innanocrystalline ferromagnets, IEEE T. Magn., 26 (1990) 1397-1402.

[105] K. Suzuki, N. Kataoka, A. Inoue, A. Makino, T. Masumoto, High Saturation Magnetization and Soft Magnetic Properties of BCC Fe--Zr--B Alloys With Ultrafine Grain Structure, Mater. Trans., JIM, 31 (1990)

743-746.

[106] K. Suzuki, M. Kikuchi, A. Makino, A. Inoue, T. Masumoto, Changes in Microstructure and Soft Magnetic-Properties of an $Fe_{86}Zr_7B_6Cu_1$ Amorphous Alloy Upon Crystallization, Mater. Trans., JIM, 32 (1991) 961-968.

[107] A. Makino, A. Inoue, T. Masumoto, Nanocrystalline soft-magnetic Fe-M-B (M=Zr, Hf, Nb) alloys produced by crystallization of amorphous phase, Mater. Trans., JIM, 36 (1995) 924-938.

[108] A. Makino, T. Hatanai, A. Inoue, T. Masumoto, Nanocrystalline soft magnetic Fe-M-B (M=Zr, Hf, Nb) alloys and their applications, Mater. Sci. Eng. A-Struct., 226 (1997) 594-602.

[109] M.A. Willard, D.E. Laughlin, M.E. McHenry, D. Thoma, K. Sickafus, J.O. Cross, V.G. Harris, Structure and magnetic properties of $(Fe_{0.5}Co_{0.5})_{88}Zr_7B_4Cu_1$ nanocrystalline alloys, J. Appl. Phys., 84 (1998) 6773-6777.

[110] M.A. Willard, M.Q. Huang, D.E. Laughlin, M.E. McHenry, J.O. Cross, V.G. Harris, C. Franchetti, Magnetic properties of HITPERM $(Fe,Co)_{88}Zr_7B_4Cu_1$ magnets, J. Appl. Phys., 85 (1999) 4421-4423.

[111] I. Škorvánek, J. Marcin, T. Krenický, J. Kováč, P. Švec, D. Janičkovič, Improved soft magnetic behaviour in field-annealed nanocrystalline Hitperm alloys, J. Magn. Magn. Mater., 304 (2006) 203-207.

[112] B.L. Shen, H. Kimura, A. Inoue, Structure and magnetic properties of Fe42.5Co42.5Nb7B8 nanocrystalline alloy, Mater. Trans., 43 (2002) 589-592.

[113] F. Kong, B. Shen, A. Makino, A. Inoue, Enhancement of soft magnetic properties of FeCoNbB nanocrystalline alloys with Cu and Ni additions, Thin Solid Films, 519 (2011) 8280-8282.

[114] M. Ohta, Y. Yoshizawa, Improvement of soft magnetic properties in $(Fe_{0.85}B_{0.15})_{(100-x)}Cu_x$ melt-spun alloys, Mater. Trans., 48 (2007) 2378-2380.

[115] M. Ohta, Y. Yoshizawa, Cu addition effect on soft magnetic properties in Fe-Si-B alloy system, J. Appl. Phys. 103, (2008) 07E722.

[116] M. Ohta, Y. Yoshizawa, Magnetic properties of nanocrystalline

$Fe_{82.65}Cu_{1.35}Si_xB_{16-x}$ alloys (x=0-7), Appl. Phys. Lett., 91 (2007) 062517.

[117] M. Ohta, Y. Yoshizawa, Effect of Heating Rate on Soft Magnetic Properties in Nanocrystalline $Fe_{80.5}Cu_{1.5}Si_4B_{14}$ and $Fe_{82}Cu_1Nb_1Si_4B_{12}$ Alloys, Appl. Phys. Express, 2 (2009) .

[118] A. Makino, M. Bingo, T. Bitoh, K. Yubuta, A. Inoue, Improvement of soft magnetic properties by simultaneous addition of P and Cu for nanocrystalline FeNbB alloys, J. Appl. Phys., 101 (2007) 09N117.

[119] A. Makino, H. Men, T. Kubota, K. Yubuta, A. Inoue, New Excellent Soft Magnetic FeSiBPCu Nanocrystallized Alloys With High B_s of 1.9 T From Nanohetero-Amorphous Phase, IEEE T. Magn., 45 (2009) 4302-4305.

[120] A. Makino, H. Men, T. Kubota, K. Yubuta, A. Inoue, FeSiBPCu Nanocrystalline Soft Magnetic Alloys with High B_s of 1.9 Tesla Produced by Crystallizing Hetero-Amorphous Phase, Mater. Trans., 50 (2009) 204-209.

[121] A. Makino, T. Kubota, K. Yubuta, A. Inoue, A. Urata, H. Matsumoto, S. Yoshida, Low core losses and magnetic properties of $Fe_{85-86}Si_{1-2}B_8P_4Cu_1$ nanocrystalline alloys with high B for power applications, J Appl Phys, 109 (2011) 07A302.

[122] A. Makino, H. Men, T. Kubota, K. Yubuta, A. Inoue, New Fe-metalloids based nanocrystalline alloys with high B_s of 1.9 T and excellent magnetic softness, J. Appl. Phys., 105 (2009) 07A308.

[123] F. Kong, A. Wang, X. Fan, H. Men, B. Shen, G. Xie, A. Makino, A. Inoue, High B_s $Fe_{84-x}Si_4B_8P_4Cu_x$ (x=0-1.5) nanocrystalline alloys with excellent magnetic softness, J. Appl. Phys., 109 (2011) 07A303.

[124] F. Kong, H. Men, T. Liu, B. Shen, Effect of P to B concentration ratio on soft magnetic properties in FeSiBPCu nanocrystalline alloys, J. Appl. Phys., 111 (2012) 07A311.

[125] X. Fan, A. Ma, H. Men, G. Xie, B. Shen, A. Makino, A. Inoue, Fe-based nanocrystalline FeBCCu soft magnetic alloys with high magnetic flux density, J. Appl. Phys., 109 (2011) 07A314.

[126] X.D. Fan, H. Men, A.B. Ma, B.L. Shen, Soft magnetic properties in

$Fe_{84-x}B_{10}C_6Cu_x$ nanocrystalline alloys, J. Magn. Magn. Mater., 326 (2013) 22-27.

[127] X. Fan, H. Men, A. Ma, B. Shen, The influence of Si substitution on soft magnetic properties and crystallization behavior in $Fe_{83}B_{10}C_{6-x}Si_xCu_1$ alloy system, Sci. China Tech. Sci., 55 (2012) 2416-2419.

[128] M. Ohta, Y. Yoshizawa, New high B_s Fe-based nanocrystalline soft magnetic alloys, JPN. J. Appl. Phys. 2, 46 (2007) L477-L479.

[129] M. Ohta, Y. Yoshizawa, Magnetic properties of high B_s Fe-Cu-Si-B nanocrystalline soft magnetic alloys, J. Magn. Magn. Mater., 320 (2008) e750-e753.

[130] M. Ohta, Y. Yoshizawa, High B_s nanocrystalline $Fe_{84-x-y}Cu_xNb_ySi_4B_{12}$ alloys (x=0.0-1.4, y=0.0-2.5), J. Magn. Magn. Mater., 321 (2009) 2220-2224.

[131] A. Makino, H. Men, K. Yubuta, T. Kubota, Soft magnetic FeSiBPCu heteroamorphous alloys with high Fe content, J. Appl. Phys., 105 (2009) 013922.